SpringerBriefs in Environment, Security, Development and Peace

Volume 19

Series editor

Dr. Hans Günter Brauch, Mosbach, Germany

For further volumes:
http://www.springer.com/series/10357
http://www.afes-press-books.de/html/SpringerBriefs_ESDP.htm

John A. Kington

Frederic W. Harmer: A Scientific Biography

Springer

University of East Anglia

John A. Kington
School of Environmental Sciences
University of East Anglia
Norwich
UK

ISSN 2193-3162 ISSN 2193-3170 (electronic)
ISBN 978-3-319-07703-1 ISBN 978-3-319-07704-8 (eBook)
DOI 10.1007/978-3-319-07704-8
Springer Cham Heidelberg New York Dordrecht London

Library of Congress Control Number: 2014941310

This photograph of Frederic William Harmer was taken (it is inferred) at his home in Cringleford, near Norwich. Harmer, appropriately, is pointing with his left hand to a mollusc shell (M. Harmer, 1998). The permission to use this photo was granted by the British Geological Survey 2013. Photographs from the Frederic William Harmer Collection are reproduced by permission of the British Geological Survey. © NERC. All rights reserved. CP13/050.

Copyediting: PD Dr. Hans Günter Brauch, AFES-PRESS e.V., Mosbach, Germany

Printed on acid-free paper

Springer is part of Springer Science+Business Media (www.springer.com)

Foreword

Frederic W. Harmer (FWH) lived in an era of all-rounder education. The educated Victorian Englishman aspired to be a polymath, competent in Classics, Mathematics, Literature, Art, Poetry, The Grand Tour, Business, Civics, Politics, Philanthropy, Domesticity, Science, and Religion. Reading their biographies and their printed works brings forth the feeling that as we moderns progress in our education, we learn more and more about less and less.

John Kington's biography and its forerunner article in the *Bulletin of the Geological Society of Norfolk* (2013) builds a picture of the 'all-rounderness' of FWH. He was a clothing manufacturer in Norwich, whose first factory was on the site of the current City Hall and whose second one is now a multi-storey car park. He served as Alderman of Norwich, as a Magistrate and for 1 year as its Mayor. He was a Congregational Church member who in 1881, as Alderman, laid the foundation stone of Shipdham Congregational Church (now United Reformed), and was a founder of the Norwich YMCA. He was a renowned amateur geologist, palaeontologist and palaeometeorologist, involved in the scientific societies of his time, later an honorary M.A. of Cambridge University. A family man, he had five children. One of his four sons, Sir Sidney, KBE, was a Professor of Zoology and Director of the London Natural History Museum, another, William, was a consultant surgeon and pioneer radiologist specialising in the treatment of throat cancer. His daughter, Edith is known to me only in a memorial note which I found in my copy of volume 2, part 3 of the *Pliocene Mollusca,* published in December 1925. It reads 'In Memory of Frederic William Harmer, M.A., F.G.S, who died peacefully on April 11, 1923, in his 88th year'.

My initiation into the personality and achievements of FWH came when I, freshly graduated in Zoology from Trinity College, Dublin, joined the Sub-Department of Quaternary Research housed in the Botany Department of the University of Cambridge.

I spent months reading the classic literature on the 'Crag' deposits of East Anglia and devising ways of extracting information from their fossil Mollusca. I made myself familiar with the works of "SVW"—S. V. Wood (senior and junior), three volumes of the *Crag Mollusca* (1848–1882) and of FWH, two volumes of the *Pliocene Mollusca of Great Britain* (1914–1925)—these books, placed side by side, take up 19 cms on my bookshelves, and weigh 8.8 kgs.

My 'Harmer' was part of FWH's household goods and contains interesting inscriptions and correspondence preceding and following his death.

Harmer's interest in geology lasted much of his lifetime, from his meeting with Searles Wood (Jnr) to SVW's death in 1884 and continued, after a furlough of almost 10 years of intense involvement in business and civic affairs, until the end of his life.

Harmer and the geologists of his time were fortunate to have access to many surface exposures of fossiliferous material, but little material was available from depth. Recent research has progressed as a result of numerous boreholes.

Many of the Victorians were primarily collectors, wanting new species and varieties for their 'Cabinets' and sometimes omitting to give details of their provenance. In some cases they tipped out from their boxes specimens whose subsequent discovery by others gave rise to spurious distribution records. Nevertheless, the sifting of many tons of material gave rise to a list of Mollusca archived in the monographs referred to above. Harmer estimated that he researched some 200 tons of material from Little Oakley. His chauffeur appears to have done a great deal of the digging. Harmer named many new species and varieties, his tendency being that of a taxonomic 'splitter' rather than a 'lumper'. I examined his *Littorina littorea* var. *distorta* from Bramerton which I found to be not varietal but pathological, owing to a clearly observable barnacle which settled on the inner lip of its aperture.

To Harmer's efforts we owe a stratigraphic division of the East Anglian Early Pleistocene into Gedgravian, Boytonian, Waltonian, Newbournian, Butleyan, *Scobicularia* Crag, Icenian, the Icenian subdivided into Norwich Horizon, Chillesford Horizon and Weybourne Horizon. The days of a malacalogically based stratigraphy are now past, despite my attempts to arrange one on the basis of palaeoecologically based Mollusc Assemblage Biozones. The emphasis in recent years has been on sedimentology. Palaeoecological investigation using Mollusc Analysis and Pollen Analysis may in future examine the various environmental facies in the lithostratigraphic units, while researches on sodium and strontium in mollusc shells yield new information on palaeosalinity and Palaeotemperature.

In times past, as one approached a school, a road sign would be encountered, depicting a flambeau torch, signifying the Flame of Learning which is passed from one generation to another. From the Victorian times into the New Elizabethan, this torch has passed from one generation of geologists to another. I express my gratitude to and admiration of those, including Harmer, from whom I took up the torch and I salute those to whom it has since passed. My gratitude is especially to John Kington, who has written this part of the Torch's Journey for us. To his readers, I wish that a flame of future research may enlighten your times also.

Wicklow, Republic of Ireland, February 2014 Peter Norton

Acknowledgments

The author is grateful to the following colleagues for their kind help and support: Peter Banham; Hans Günter Brauch; Peter Brimblecombe; Stanley Cornford; Paul Crofts; Gaston R. Demarée; Peter Humphries; Kathy Maxwell; Peter E. P. Norton; Mike Salmon; Malcolm Walker; and Tim Holt-Wilson.

British Geological Survey 2013. Photographs from the Frederic William Harmer Collection are reproduced by permission of the British Geological Survey.

Finally, my sincere thanks to my dear wife, Beryl for her devoted interest and encouraging support during all stages of the work.

Norwich, February 2014 John A. Kington

Contents

Chapter 1
Introduction

Harmer (1835–1923) was one of the pioneers in the field of East Anglian geology as well as one of the last members of a distinguished group of amateur geologists who had been responsible for making major advances in the science during the Victorian era and early years of the 20th century. In particular, he played a key role in elucidating the Pliocene and Pleistocene stratigraphy in the east of England by developing the use of mollusca for biostratigraphic correlation within the Crags of East Anglia.[1]

This book, comprising the first definitive account of the geological and palaeo-meteorological studies made by Harmer, contributes a previously missing chapter to the history of science. The main objective is to ensure that the scientific work of Harmer, which unfortunately has been widely neglected or forgotten, becomes more generally known and acknowledged. The balance of this deficiency will be redressed by bringing to light in this volume his contributions to the history of science to an audience of academic and lay readers of the current literature.

It is regrettable that no references are given to Harmer's research on the Crags of East Anglia in the current literature including *Geology of England and Wales*, Geological Society, London, 1992, and *British Regional Geology: London and the Thames Valley*, H.M.S.O., London, 1996. He was certainly a well-respected scientist in his day, being awarded Hon. M.A. Cantab., elected F.G.S. and F.R.Met.S., and a Membre Hon. de la Société Belge de Géologie et de Paléontologie. From an early age he had been an active member of the Geological Society of Norwich. In 1902 he was awarded the Murchison Medal for that year in recognition of his work on the Pliocene and Pleistocene deposits of East Anglia. A leading exponent of 19th century geological literature, Horace B. Woodward, F.R.S., F.G.S., included Harmer in a group of geologists who 'have added greatly to our knowledge of the structure of the [East Anglian Crag] deposits⋯' (Woodward 1887).

[1] This text is based on Harmer's key publications and related material and is given in Roman type; direct quotations by Harmer et al. are indented.

J. A. Kington, *Frederic W. Harmer: A Scientific Biography*, SpringerBriefs in Environment, Security, Development and Peace 19, DOI: 10.1007/978-3-319-07704-8_1,

Although at first, due to his business commitments, he only had a limited amount of leisure time to pursue his love of geology, a chance meeting on the beach at Mundesley, Norfolk with the renowned geologist, Searles V. Wood, Junior, was the beginning of a friendship and productive partnership during which, for example, they prepared a contoured-map of the glacial deposits of Norfolk and Suffolk on a scale of 1 in. to 1 mile, said to be the first ‘drift’ map of its kind in the literature.

Sadly, after the publication of further material on the Pleistocene deposits of eastern England, Wood died in 1884. The loss of his friend and co-worker deeply affected Harmer and for a time he withdrew from active geological studies and devoted himself instead to municipal duties and the politics of the day. However, on discovering, after a gap of about 10 years, that his views had found a sympathetic hearing by a new generation of geologists, he resumed an intensive study of the Tertiary and Quaternary deposits, both in eastern England and related parts of mainland Europe. This resulted in a series of papers on the East Anglian Crags, which as standard references for the subject, inaugurated a new chapter in the geology of the region. His contributions to glaciology and palaeometeorology during this later period were no less innovative. His studies as a pioneer palaeometeorologist in which he advocated the reconstruction of past circulation patterns comprises an integral part of this volume.

Harmer developed a close relationship with many leading scientists of his time, both in Britain and mainland Europe and in 1907 he organised a field excursion to Norfolk in order to examine the geology of the county under his expert guidance. For this event he invited a number of renowned geologists and related earth scientists from the international community and their willingness to accept and attend this excursion reflects the high esteem by which Harmer was regarded at that time amongst his peers, both at home and abroad as will be described in more detail in Chap. 3.

Furthermore in journeys to mainland Europe, Harmer visited colleagues in France, Belgium, Holland, Denmark, Italy and Switzerland to discuss and examine fossiliferous specimens, such as the large collection assembled by the Belgian malacologist, Philippe Dautzenberg at the Royal Belgian Institute of Natural Sciences in Brussels.

During the final years of his life Harmer produced two outstanding works: firstly, a contoured map showing the distribution of erratics and drift in England and Wales (posthumously published, Harmer 1928); and secondly his *magnum opus*, a two-volume monograph, *The Pliocene Mollusca of Great Britain* (Harmer 1914–1925); this latter work remains a standard reference on the subject in which he had meticulously examined and copiously illustrated the rich deposits of fossilized shellfish in Britain, especially those in northeast Essex and East Anglia. This massive achievement provides a fitting monument to his outstanding work as a geologist.

References

Harmer, F.W., 1914–1919: *A Monograph of the Pliocene Mollusca of Great Britain*, being supplementary to Searles V. Wood's *Monograph of the Crag Mollusca*, Volume I (London: Palaeontographical Society).

Harmer, F.W., 1920–1925: *A Monograph of the Pliocene Mollusca of Great Britain*, being supplementary to Searles V. Wood's *Monograph of the Crag Mollusca*, Volume II, (London: Palaeontographical Society).

Harmer, F.W., 1928 (posthumously): "The Distribution of Erratics and Drift", in: *Proceedings of the Yorkshire Geological Society*, 21: 79–150.

Woodward, H.B., 1887: *The Geology of England and Wales: With Notes on the Physical Features of the Country*, Second Edn. (London: George Philip & Son).

Chapter 2
Family Background and Public Life

Abstract Details are given about Harmer's family background and public life, including his education; homes in Norwich and Cringleford; marriage to Mary Lyon and their five children, John, Sidney, Edith, Thomas and William, John and Thomas following their father into the family business, Sidney becoming a zoologist and William a consultant surgeon; Harmer's career in the family clothing business; interests in geology and palaeometeorology; membership of various geological societies and the Royal Meteorological Society; scientific awards; public duties in Norwich, Alderman (1880), Magistrate (1884) and Mayor (1887-1888); membership of two church congregations, first the Congregational Church, Norwich and later St Peter's, Parish Church, Cringleford. Born in Norwich in 1835 he died in Cringleford in 1923, aged 87; correspondence with Charles Darwin.

Keywords Family · Homes · Children · Clothing business · Public affairs · Church matters · Geological and palaeometeorological interests

2.1 Introduction

Esse quam videri, 'To be, rather than to seem (to be)', from the essay, *On Friendship* by the Roman statesman, Cicero, was the motto adopted by Harmer's family; this maxim provided a guiding principle that Harmer followed during his long and productive life.

J. A. Kington, *Frederic W. Harmer: A Scientific Biography*, SpringerBriefs in Environment, Security, Development and Peace 19, DOI: 10.1007/978-3-319-07704-8_2,

Harmer was a descendant of a long-established Norfolk family. Born in Norwich on 24 April 1835 he died in Cringleford, near Norwich on 11 April 1923, aged 87. Harmer was privately educated and in 1849 was awarded a "School Certificate" for 'Progress in General Knowledge'.

At the age of 15 he joined the family business in Norwich (one of the oldest clothing manufacturing firms in the country) and began work at the firm as a clerk to his father, Thomas. At this time Harmer lived with his parents at 1 Newmarket Terrace (since re-named Albert Terrace) off the Newmarket Road, Norwich. In due course, he was appointed director of the firm.

In 1856 Harmer became a member of the Congregational Church in Princes Street, Norwich. Four years later he married Mary (third daughter of Adam Lyon of Downham, Cambridge) and they moved into a rented house at Grove Place, Heigham in Norwich. They had five children, all born in Heigham, namely, John Alexander (born 27 October 1860), Sidney Frederic (born 9 March 1862), Edith Mary (born 5 October 1864), Thomas Bertrand (born 15 November 1867) and William Douglas (born 25 August 1873).

The eldest son, John, joined the firm in 1879 and by assisting with the running of the business gave his father more time to pursue his interests in geology and public affairs. John who lived at Boyton House, Ipswich Road, Norwich became a co-partner of the firm in 1897; he died on 17 November 1925, aged 65 and, like his father, was buried at St Peter's Church, Cringleford.

The second son, Sidney, became a distinguished zoologist, President of the Linnaean Society (1927–1931) and was awarded the Linnaean Medal in 1934. He was also Director of the Natural History Department, British Museum from 1919 until his retirement in 1927. Sidney lived at *The Old Manor House*, Melbourn, near Royston in Hertfordshire. He was invested as a Fellow, Royal Society (F.R.S.) and, in 1920, a Knight Commander, Order of the British Empire (K.B.E.). Sidney died on 22 October 1950, aged 88. Sidney's son, Russell Thomas Harmer, joined the firm in 1919, became a partner in 1925 and was the 5th generation to be involved in the business.

Harmer's only daughter, Edith, did not marry and in 1923, freed of her duties at *Oakland House* on the death of her father, moved to France where she lived for the rest of her life in Paris. The third son, Thomas joined the firm in 1888 and, like his brother John, became a co-partner in 1897. The youngest son, William, became a consulting surgeon at St Bartholomew's Hospital in London, specialising in diseases of the throat and nose. He was a pioneer in the treatment of throat cancer by radium at the Radium Institute, London. During the Second World War his son, Squadron Leader Michael Harmer, was in charge of surgical wards at the R.A.F. Hospital, Hoylake. William died in 1962, aged 89 (Fig. 2.1).

Fig. 2.1 Frederic W. Harmer, Mayor of Norwich, 1887–1888 (Norton 2013)

Besides being a successful businessman Harmer was scholarly and well-travelled, making repeated journeys to mainland Europe where he was able to visit many geological museums in Denmark, Belgium, France and Italy. His scientific lifestyle was maintained by the success of the family clothing business. However, Harmer was also an outstandingly public-spirited figure who played an active role in Norwich affairs, serving as an Alderman in 1880, a Magistrate in 1884 and Mayor in 1887–1888.

None of this meant that his interest in the firm had lessened in any way and not long after his Mayoralty, Harmer, realising the need for more spacious factory premises, decided that as the original factory in Bethel Street was becoming too small for the expanding business bought a much larger site in St Andrew's Street.

As with the building of his home in Cringleford, *Oakham House* (see below), Harmer turned to the architectural skills of the Boardman family, this time with the son of Edward Boardman, Harmer's friend and fellow church member. Having recently qualified as an architect, the younger Boardman joined his father's business in Norwich and drew up plans for the new factory which was opened in St Andrew's Street in 1890. Known as St Andrew's Steam Clothing & Hosiery Works, it became the main base of the firm's future activities until it was severely damaged by bombing during an air raid in 1943.

During the late 19th century several large houses were built in Cringleford by wealthy Norwich businessmen who wished to move out of the city into a more rural setting. In about 1872–1873, Harmer purchased about sixteen acres of land adjoining Newfound Lane, now known as Colney Lane, Cringleford. He commissioned Edward Boardman, the City architect, to draw up plans for the erection of a large house on this site.

In 1876, Harmer and his family moved from their former home in Heigham to the newly-built *Oakland House* in Cringleford (home village, incidentally, of the present writer) (Fig. 2.2). Domestic and ground staff were engaged, the former usually numbered about five and the latter included eight gardeners. In the same year, a small, single storey lodge was built at the entrance to the drive on Colney Lane; this building still stands today in an extended form. A coachman's house built in the same year also stands in Colney Lane. Harmer kept a number of horse-drawn carriages and the sale catalogue for the auction of family possessions after his death listed a Morris Brougham, a Landau, a Victoria and a shooting cart. Harmer's will also indicated that he owned a motor car which he used extensively on his geological expeditions in the early 1900s.

Fig. 2.2 *Oakland House*, Cringleford, near Norwich (Cringleford Historical Society 2006)

Harmer was one of the founders of the Norwich Electricity Company. He was also a founder of the Norwich Young Men's Christian Association (YMCA) and was, for some years, its vice-president. He was a trustee of the Norwich Grammar School, now known as the King Edward VI, Norwich School. Harmer and Sir Peter Eade, M.D., F.R.C.P. (Mayor of Norwich 1883, 1893 and 1895) were jointly responsible for the layout of Chapel Field Gardens, the gardens surrounding Norwich Castle and the acquisition of Mousehold Heath.

Fig. 2.3 Frederic W. Harmer awarded Honorary M.A., University of Cambridge, 1918 (Norton 2013)

From a young age Harmer was an active member of the Norwich Geological Society and in 1864 met the geologist Searles V. Wood Jun. (1830–1884). Harmer was a founder member of the Norfolk & Norwich Naturalists' Society, was its president 1877–1879 and subsequently a vice-president. He was also a member of the Norfolk & Norwich Horticultural Society.

He was a member of the council and elected a Fellow of the Geological Society of London, 1896–1900; awarded the Murchison Medal, 1902; a member of the council of the Palaeontographical society, 1878–1882 and 1905–1906 and in 1918 was awarded an Honorary M.A. by the University of Cambridge (Fig. 2.3). He was also a member of several geological societies in mainland Europe.

Harmer became an authority on the geology of East Anglia. Through his geological studies he became a pioneer of palaeometeorology and was elected a Fellow of the Royal Meteorological Society.

Following their move from the city to Cringleford, Harmer and his wife joined the congregation of the Parish Church, St Peter's where, for a number of years, he acted as honorary organist—a man truly of many talents. Harmer also took an active role in the village life of Cringleford and was chairman of the Parish Council for 20 years. He was president of the Patteson Club in the village and was probably instrumental in the Cringleford Horticultural Society coming into existence as a section of the Patteson Club. In the 1920s, it was customary for the Annual Show of the Society to be held in a garden of one of the larger properties. Harmer gave permission for the third show to be held in the grounds of *Oakland House*. This event was planned to be more ambitious than on previous occasions with the Red Cross Band being engaged to play, and tents and tables hired for the Bank Holiday Monday but, unfortunately, it apparently poured with rain during the afternoon!.

Behind the house sloping down to the river Yare were several acres of open ground which Harmer developed into a large walled garden containing a conservatory, greenhouse and vinery. The central feature of the garden, known as 'the park', was an avenue of ancient oak trees after which the property was named. The 'wild' part of the garden was particularly beautiful in the spring when massed clusters of flowers such as snow flakes and daffodils were in bloom. There was also a water garden with an iron gate leading onto a bridge: a water feature in this part was an important and practical ornament since it provided ice for the icehouse. Alongside the river there were water meadows and a wild waterlogged alder carr which Frederic drained and planted with bulbs (Fig. 2.4).

Fig. 2.4 Inferred view of Harmer's garden at *Oakland House*, Cringleford, near Norwich (P680279, CP13/050. Reproduced by permission of the British Geological Survey © NERC. All rights reserved)

Harmer's gardens would have provided a most beautiful setting for his home; however the area today is covered with a maze of houses and roads.

2.2 Charles Darwin

A fascinating connection between Harmer and the eminent naturalist Charles Darwin (1809–1882) came to light when a signed original letter from Darwin to Harmer was recently discovered in the archives of the Ipswich Museum, Suffolk.

Down Beckenham
Kent
August 29th 1872

Dear Sir
I am very sorry that you have become involved in a troublesome controversy on my part. The sentence given by Mr Lyon in inverted commas is an invention, and it is a most unjustifiable proceeding on his part. He might of course have given any interpretation which he pleased of my words, but he had no right to put the words in inverted commas. I may add that I have given in the later and more especially in the 6th Ed. Of the Origin many cases showing how common generated forms (that is forms partaking of the characters of existing defined groups) are in all the same ancient formations. I have always been curious to hear who wrote Homo Vs. Darwin. Mr Lyon can hardly have the disposition of a gentleman, for in one place he states that I speak the truth solely because I should be found out if I lied. I hope that you will soon be able to bring your controversy to an end and I remain in haste.

Dear Sir
Yours faithfully
Ch. Darwin

It is believed that this letter is in reply to one written to Darwin by Harmer who apparently had entered into a newspaper controversy with W.P. Lyon, author of the publication 'Homo versus Darwin' published in 1872, in which Lyon (an Independent Minister from Tunbridge Wells) ascribed to Darwin the saying: 'natural selection is a kind of god that never slumbers or sleeps'.

Harmer did not believe that Darwin had made this statement and had written to Darwin to this effect.

In the letter housed in the Ipswich Museum Darwin is interested to learn that Lyon is the author of 'Homo versus Darwin' which was subtitled 'A judicial examination of statements recently published by Mr Darwin regarding the Descent of Man'. This was originally published anonymously in the style of a court case transcript but was apparently based on actual correspondence between Lyon and Harmer.

In a reply to Darwin's letter of 31 August, Harmer explains that he was sucked into the controversy when he wrote a letter to the *Daily Press* (possibly a Norwich paper) replying to a published letter by a Dr Bateman of Norwich, who claimed to have proved the 'fallacy of Darwinism'.

Bateman called in Lyon to support him and at first Harmer wanted nothing more to do with Lyon. This was seen by Lyon as a collapse of Harmer's case and Harmer could then not help being drawn into the controversy.

Again in his letter of 31 August, Harmer, himself a religious man, apologises for the attacks Darwin had received and continued to receive in the name of religion by its 'self constituted champions' (Friends of the Ipswich museums 2013).

This incident illustrates how well informed Harmer was about scientific matters of his day and his active response in dealing with any ill-judged comments.

2.3 Death of Harmer

Following the death of Harmer on 11 April 1923, the proceeds of his property were left, after the payment of certain bequests and expenses, to his five children in equal shares. Edith was given first choice of her father's furniture and personal effects, paying a price settled by a valuer. Subject to her first choice, the four sons were also able to purchase similar items on the same basis. Harmer bequeathed £2,000 each to Edith, Sidney and William, his three children who had not become his co-partners in the family business.

Harmer bequeathed his scientific collections to his son, Sidney. He also left a modest annuity to his friend, Andrew Bell, for his help with the publication of scientific work and the care and distribution of natural history collections and scientific books.

Sidney replied to a letter of sympathy from the eminent geologist, Arthur Woodward, F.R.S.

> 30, Courtfield Gardens,
> S.W. 5
> April 18, 1923
>
> My dear Woodward,
> Will you accept my cordial thanks for your sympathetic letter. In his later years my father had been cut off from his scientific friends, but it was always a satisfaction to him to know that his work was appreciated. He had often been indebted to you (as we all have) for assistance in bibliographical matters, and he often spoke to me of the help he obtained from you when he wanted it. His work was practically completed, because although there is still something to be published by the Palaeontographical Society he had passed for press the Part which is to appear in due course, and he had completed the MSS, with one Plate, of what he had intended to be actually the final instalment. His scientific work was a great resource to him, and he was engaged with it almost to the end.
>
> Yours very sincerely,
> Sidney F. Harmer

References

Friends of the Ipswich Museums, 2013: Darwin's letter, at: http://www.foim.org.uk/index
Norton, P.E.P., 2013: Personal Communication.

Chapter 3
Geology

Abstract Reviews of the Pliocene and Pleistocene Epochs in East Anglia; reference to the discovery and importance of molluscan shells; account of the Crag beds including the Red Crag, Norwich Crag, Chillesford and Cromer Forest Beds and Weybourne Crag; the Crag Sea and its changing molluscan fauna in relation to the impending advance of the Scandinavian ice sheet during the late Pliocene; features of lowland glaciation including boulder-clay, moraines, erratics, pro-glacial lakes; advance and retreat of ice sheets; post-glacial times including the climatic optimum; research and field work with S.V. Wood, Jun.

Keywords Pliocene · Pleistocene · Glacial geology · East Anglian crags · Molluscs · Boulder-clay · Moraines · Erratics · Ice sheets

3.1 General

East Anglia is a region unmatched in the British Isles for the study of lowland glaciation and due to his Norfolk family roots and a predilection for geology, Harmer, from an early age, devoted his leisure time to the study of glacial landforms in his home region. Later, when two of his sons took over the responsibility for the day-to-day running of his clothing business, geological studies became Harmer's main preoccupation. As a result of his meticulous research and outstanding field work undertaken over many years, Harmer became known in the scientific community as a leading authority of the Pliocene and Pleistocene deposits that widely occur in East Anglia.

J. A. Kington, *Frederic W. Harmer: A Scientific Biography*, SpringerBriefs in Environment, Security, Development and Peace 19, DOI: 10.1007/978-3-319-07704-8_3,

3.2 Geological Classification

Although the research carried out by Harmer on the East Anglian Crags presented in this book employ geological terms that were in use during his time, the stratigraphy of the Red Crag and the Norwich Crag have now been assigned to the Pleistocene rather than Pliocene Epoch. In 1948 the International Geological Congress recommended that the base of the Red Crag should be adopted as the beginning of the Pleistocene Epoch in Britain, with only the Coralline Crag remaining in the Pliocene. Further criteria for placing the Pliocene-Pleistocene boundary between the Coralline and Red Crags, including the presence of a stratigraphical break, marked increase in the proportion of northern forms of marine molluscs (indicative of a refrigeration of the climate) and the first arrival of elephant and horse were presented in 1957 (West 1957). However, the former classification, which included all of the East Anglian Crags within the Pliocene, is employed in this volume in order to retain continuity with the geological studies undertaken by Harmer.

3.2.1 Introduction

Harmer was very much a 'hands-on' geologist who based his research on extensive studies in the field. A series of unpublished notebooks dating from 1865 provide evidence of Harmer's activities in geological field work; further notebooks contain fragments from Searles V. Wood, Junior's field work and references to Geological Survey Memoirs. Other unpublished material illustrates Harmer's wide-ranging interests in earth science such as the nomenclature of geological formations, a lecture on the geology of Norfolk, Pleistocene deposits of East Anglia; miscellaneous notes on glacial geology, boulders and various borings; notes by Professor Kendall; miscellaneous survey memoirs referring to occurrences of boulder-clay; notes and discussion of a paper by Professor P.G.H. Boswell; notes on the Crag formations; notes on the Cromer Forest Bed Series of East Anglia; and letters to fellow geologists such as J. Geikie, Kitchen, Allen and H.B. Woodward.

3.2.2 Pliocene and Pleistocene Deposits of East Anglia

Harmer made many field excursions to geological sites in East Anglia and was never happier than when demonstrating in situ the varied deposits of the region to interested groups. For example, on 9–14 July 1903 he led a field excursion to Cromer, Norwich and Lowestoft. In an account published later in the *Proceedings of the Yorkshire Geological Society* he wrote:

> The special object of this excursion was to study, as far as might be possible during the short time allotted to it, the glacial deposits of Norfolk and their relation to those of the North of England.
>
> Assembling the night before at Tucker's Royal Hotel, Cromer, the excursionists took train to North Walsham and on the morning of Thursday, July 9th, proceeding thence by carriage to Hasboro' (Happisburgh), at the south-eastern termination of the cliff section. A pleasant walk of about eight miles along the beach, in delightful weather, which fortunately lasted during the whole visit, brought them to Trimingham; there a conveyance awaited them for North Walsham, from whence they returned to Cromer in the evening.
>
> The Pleistocene beds of East Anglia were divided by the late Searles V. Wood, Jun., into Lower, Middle, and Upper, those of the North Norfolk coast, the Cromer Till and the Contorted Drift, being placed by him in the first division. The Cromer Till, a bed of tough unstratified boulder-clay, of a dark blue colour, occurs at Hasboro' at the base of the cliff, there about 30 feet [9 metres] in height, it was seen to contain abundantly fragments of grey flint and hard chalk from the Wolds of Lincolnshire or Yorkshire, some of the chalk being scratched or striated, together with many broken shells of recent species, especially Tellina balthica [Baltic tellin, a widely distributed mollusc, its tolerance of low salinities explains its presence in the Baltic and hence its name], Cardium edule [Common cockle, a widely distributed mollusc] and Cyprina islandica [one of the largest bivalve molluscs of British waters] (Harmer 1904).

Another notable excursion was made in 1907, to which he had invited leading geologists from both Britain and mainland Europe, is especially interesting as it is illustrated with photographs from the 'Frederic William Harmer Collection', presently held in the archives of the British Geological Survey (British Geological Survey 2013). Harmer's residence in Cringleford, *Oakland House* would have been large enough to accommodate at least some of the participants including, perhaps, one of the leading invitees, Professor Albrecht Penck.

The participants included: Professor Knut O. Bjørlykke, Norwegian geologist; Dr Marcellin Boule, French palaeontologist; Professor Carl H. Credner, German geologist; Professor Erich von Drygalski, German geographer; Dr Carl C. Gottsche, German geologist; Professor Thomas M. Hughes, Welsh geologist; Dr Manson, English geologist; Professor Albrecht Penck, German geologist and geographer; Dr Robert H. Rastall, English geologist; Dr Jakob Sederholm, Finnish geologist; Dr Knud J.V. Steenstrup, Danish geologist; Dr Emil E. A. Tietze, Austrian geologist; Professor Niels V. Ussing, Danish scientist; and Dr W. Van der Gracht, Dutch geologist.

A photograph album containing pictures which had been taken on this excursion was later presented to Harmer by Penck with the following dedication: 'A souvenir of a beautiful excursion in Norfolk given to his excellent guide Mr Harmer. Albrecht Penck, Berlin August 1911' (Figs. 3.1 and 3.2).

Fig. 3.1 P680263 (1907) Albrecht Penck standing on a rock-strewn moraine deposited by a former Scandinavian ice sheet, probably near Utrecht, Holland. (CP13/050 Reproduced by permission of the British Geological Survey © NERC.

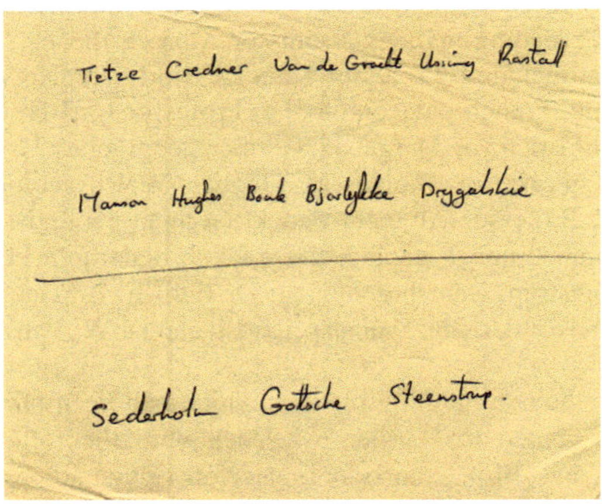

Tietze Credner Van de Gracht Ussing Rastall

Manson Hughes Boule Bjørlykke Drygalskie

Sederholm Gottsche Steenstrup.

Fig. 3.2 P680277 (1907) Harmer's geological excursion to Norfolk. A sheet of paper which was attached to images P680275 and P680276 with the names of some of the scientists who had been invited by Harmer to attend the 1907 excursion: Professor Knut O. Bjørlykke, Norwegian geologist; Dr Marcellin Boule, French palaeontologist; Professor Carl H. Credner, German geologist; Professor Erich von Drygalski, German geographer; Dr Carl C. Gottsche, German geologist; Professor Thomas M. Hughes, Welsh geologist; Dr Manson, English geologist (?); Dr Robert H. Rastall, English geologist; Dr Jakob Sederholm, Finnish geologist; Dr Knud J.V. Steenstrup, Danish geologist; Dr Emil E. A. Tietze, Austrian geologist; Professor Niels V. Ussing, Danish scientist; and Dr W. Van der Gracht, Dutch geologist. (CP13/050 Reproduced by permission of the British Geological Survey © NERC.

Interestingly, Penck does not appear in any of the pictures which suggests that he was the photographer who, on his return to Berlin, arranged to have the photographs bound in an album. This gift from Penck who, with the Austrian geographer Eduard Brückner, was engaged at the time on the classic four-fold glacial sequence regarding the advance and retreat of Alpine glaciers, *Die Alpen im Eiszeitalter* (1901–1909), would have been a treasured possession of Harmer (Figs. 3.3, 3.4 and 3.5).

Fig. 3.3 P680275 (1907) Harmer's geological excursion to Norfolk, participants at Sheringham (Harmer on *left*) (CP13/050 Reproduced by permission of the British Geological Survey © NERC. All rights reserved)

Fig. 3.4 P680276 (1907) Harmer's geological excursion to Norfolk; participants at Sheringham (Harmer third *left*) (CP13/050 Reproduced by permission of the British Geological Survey © NERC. All rights reserved)

Fig. 3.5 P680262 (1907) Harmer's geological excursion to Norfolk some participants resting on a shingle beach looking north towards Lowestoft, Suffolk (CP13/050 Reproduced by permission of the British Geological Survey © NERC. All rights reserved)

In 1910 Harmer was invited to contribute chapters on the Pliocene and Pleistocene Epochs in eastern England for the Jubilee Volume of the Geologists Association, *Geology in the Field* (1858–1908). The following accounts are based on these chapters together with other key works.

3.3 The Pliocene Epoch in East Anglia

During the Pliocene Epoch the British Isles was shaped by various geological forces into a physical pattern which more or less defines the geography of the region today. However, there were some regional differences. For example, at the beginning of the period southeastern England was about 200 m (650 ft.) lower than today and the crest of the Wealden dome would have formed a low island from which the former surface of chalk had been removed to expose the underlying rocks; further north in eastern England the shore-line of the Pliocene or Crag Sea would have extended farther west than that of the present North Sea. The Crag Sea was essentially a great gulf into which the proto-Rhine flowed to be joined on its left bank by the proto-Thames; the deposits of this river system contributed to the formation of the Pliocene Crag. Whilst initially the Crag Sea lingered in the syncline of what today is the London Basin, it later progressively retreated northeast to occupy the position of the present North Sea.

3.3.1 Pliocene Crag

In East Anglia, the Pliocene Crag beds, ranging in age from the Coralline to the Weybourne deposits (oldest to youngest), were of particular interest to Harmer as they represented successive stages during a period when tectonic changes were accompanied by a refrigeration of the climate. Coralline Crag, the oldest of these deposits comprises fragmentary polyzoa, sea-mats and related species, formerly termed corallines. As the land rose and the Crag Sea retreated northeast, later deposits are not superimposed on earlier beds but are found progressively further away from the original Pliocene basin.

Immense numbers of molluscan shells were discovered in the Crag beds with the older shells generally of warm-water species, indicating that in earlier times a seaway was probably open through central Europe to warmer waters. However with the cooling of the climate during the Pliocene epoch there was a gradual decline and eventual disappearance of the southern forms of molluscs of the earlier zones with the shells of northern species becoming steadily more and more important.

3.3.2 The Red Crag

Succeeding the Coralline Crag in the geological table, the Red Crag, comprising quartz-rich sandstone stained orange-red with iron oxides, lies unconformably above the Eocene London Clay and was deposited a short distance offshore in a warm and relatively shallow sea about 2–3 million years ago. Its rich content of mostly broken and abraded fossilised shells indicates a highly energetic marine and atmospheric environment in which material from the sea bed was repeatedly swept up onto contiguous shorelines due to strong winds and associated wave action. During this period the molluscan shells of a northern origin become increasingly more numerous at the expense of southern ones. The Red Crag deposits can be traced along the Suffolk coast from Walton-on-the Naze to Aldeburgh with the Norwich Crag deposits occurring abruptly north of the latter site.

The earliest scientific account of fossils contained in the Red Crag was made by the naturalist-geologist, Samuel Dale (1659–1739) in the appendix of the book, *The History and Antiquities of Harwich and Dovercourt*, first collected by Silas Taylor, and now much enlarged, with notes and observations relating to natural history, London, 1730 (Challinor 1971). Dale based his account on a letter to the *Philosophical Transactions of the Royal Society* in which he commented on the yet to be named Red Crag formation as follows:

> The same strata of sand, and fragments of shells, with the same fossils imbedded [the Red Crag], are to be found at Walton Ness on the other side of the Æstuarium [Dovercourt Bay], which is 5 or 6 miles broad from Harwich, as likewise at Bawdsey Cliff in Suffolk, which is 8 or 9 miles distant, and in other cliffs on that shore, where I have met with them (Dale 1704–1705).

This observation indicated that Dale was aware of notable exposures of the Red Crag in the cliffs of northeast Essex and south Suffolk, later investigated by Harmer (Figs. 3.6 and 3.7).

Dale included two plates of the Harwich Cliff in his book and four figures illustrating fossils he had described in the text. In the preface Dale acknowledged that the basis of the book had been written by Taylor in about 1676. Following Taylor's death in Harwich 2 years later, Dale obtained and edited the manuscript with a view to its publication. Although the text, comprising 255 quarto pages, was

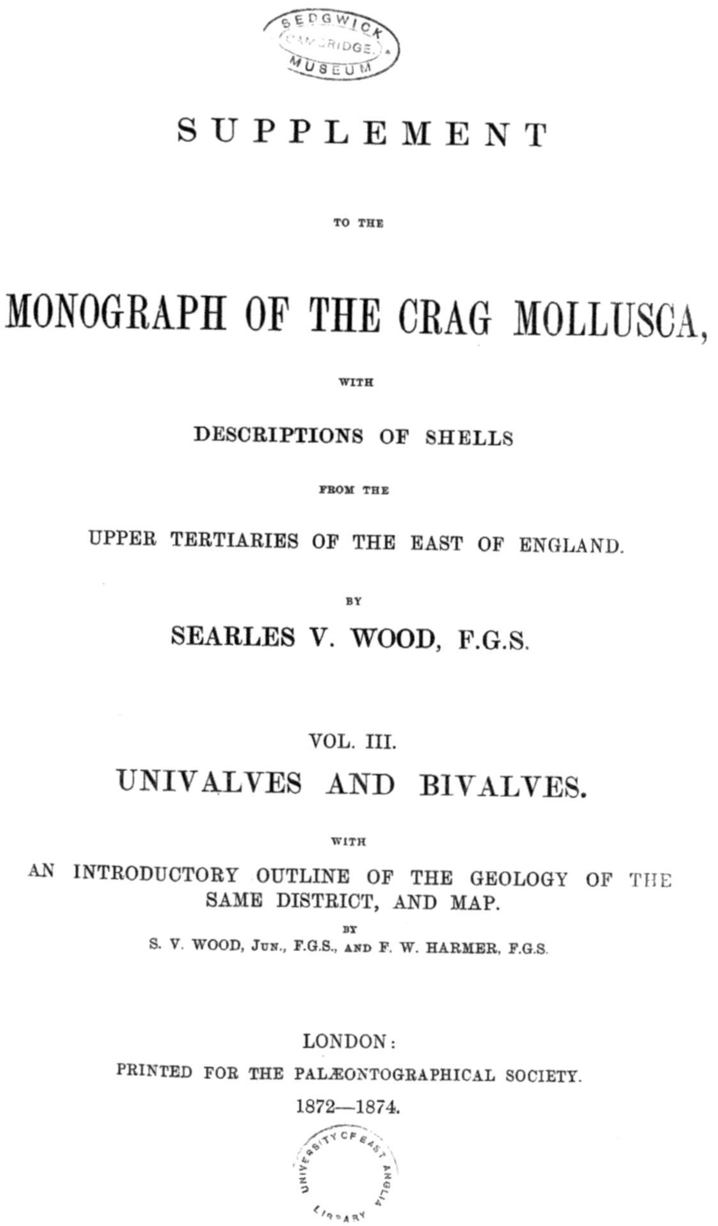

SUPPLEMENT

TO THE

MONOGRAPH OF THE CRAG MOLLUSCA,

WITH

DESCRIPTIONS OF SHELLS

FROM THE

UPPER TERTIARIES OF THE EAST OF ENGLAND.

BY

SEARLES V. WOOD, F.G.S.

VOL. III.

UNIVALVES AND BIVALVES.

WITH

AN INTRODUCTORY OUTLINE OF THE GEOLOGY OF THE SAME DISTRICT, AND MAP.

BY

S. V. WOOD, Jun., F.G.S., and F. W. HARMER, F.G.S.

LONDON:

PRINTED FOR THE PALÆONTOGRAPHICAL SOCIETY.

1872—1874.

Fig. 3.6 Title page of the Supplement to the *Monograph of the Crag Mollusca* by Searles V. Wood, Palaeontographical Society, 1872–1874, London (Wood Sen. 1872–1874)

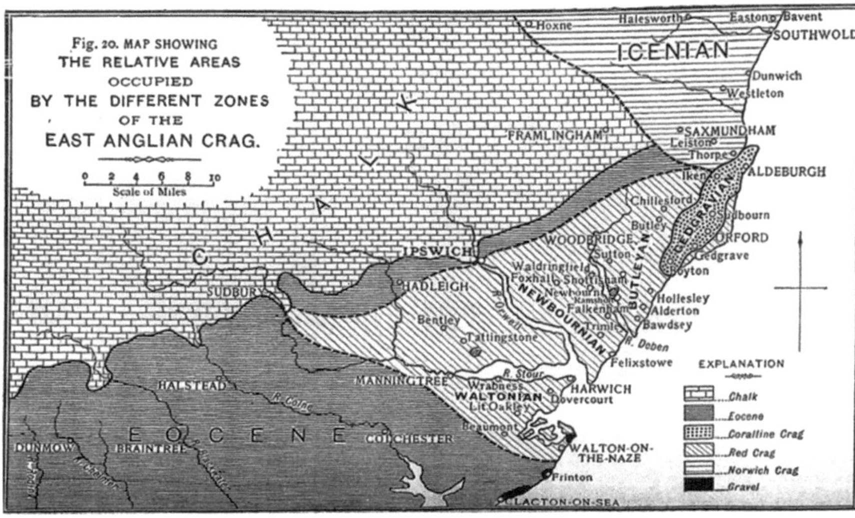

Fig. 3.7 Map showing the Pliocene Crag areas of East Anglia (Harmer 1910b)

compiled directly from Taylor's manuscript, the appendix was entirely the work of Dale and contained his account of the geology of the Harwich Cliff which he described as 'a sort of promontory, which divides Orwell Haven from the Æstuarium (Dovercourt Bay) contained between that and Walton Naze or Ness'. Over 40 fossils from the Harwich Cliff described by Dale have now been classified under the Red Crag formation according to the current terminology.

Dale also discussed the problem about the origin of fossils which had puzzled oryctologists for many years. For example, whilst the palaeontologist, Robert Plot (1640–1696) held the view that they were *lapides sui generic*, produced by some plastic power inherent in the crust of the earth, Robert Hooke (1635–1703) presciently believed that they owed their form and figuration to the shells of certain shellfish and believed that such fossils provided reliable clues to the past history of life on earth. Despite the objections of contemporary naturalists such as John Ray (1627–1705) who found the concept of extinction theologically unacceptable; Hooke stated that in certain cases they might represent species that had become extinct through some geological disaster.

Dale's account of the Red Crag fossils in the Harwich Cliff is especially valuable as it preserves a record of the contents of a former outlier of the Essex coast which was later lost to the sea. His name is preserved by a species of fossil shellfish, *Buccinum dalei* (Dale's whelk: *Liomesus dalei*). In naming it, James Sowerby (1757–1822), himself a student of the crag fossils, said he did so 'to commemorate the labours of Dale, who appears to be the first person that took notice of the Suffolk [sic] crag fossils'.

Following Dale's lead, Harmer found that much of the Red Crag was bedded at a more or less constant angle of 30°, as, for example, in the cliff section at Walton-on-the-Naze. Harmer further discovered that the beds never attain any great thickness, generally no more than about 6 m (20 ft.), but constantly present a highly inclined pattern of bedding. Such stratification indicated, as pointed out by Wood, Jun., a deposit formed against a beach, or foreshore, or on the edges of a shoal (Fig. 3.8).

Fig. 3.8 P680271 (1907) Red Crag pit at Bentley, Suffolk with the British geologist Percy G.H. Boswell (later Professor of Geology, Imperial College, London) and colleague sieving deposits. (CP13/050 Reproduced by permission of the British Geological Survey © NERC. All rights reserved)

Harmer's meeting by chance with the geologist, Searles Valentine Wood, Junior, 1830–1884 (son of the renowned palaeontologist, Wood Senior 1798–1880) on Mundesley beach and their subsequent friendship proved to be most beneficial to Harmer by enhancing his geological studies. For example, in the early 1870s they jointly presented a detailed account of the Red Crag:

> The physical structure of the Red Crag is unlike that of any other formation known to us, ancient or modern. The more considerable portion of it is formed of a succession of beds, varying from two or three, to nearly twenty feet in thickness, each of which consists of laminae of sand and shells, inclined at a high degree to the horizon. This structure is altogether different from the well known one of false bedding, which also exists in some parts of the Red Crag, especially in that under which the phosphatic nodules are worked, and the two forms of bedding pass more or less into each other. This oblique lamination may be traced (as, for instance, in Bawdsey Cliff) for a considerable distance in a constant manner, without shading off into horizontal stratification or passing into false bedding (Wood Jun. and Harmer 1872–1874).

In providing one of the best exposures of the Red Crag in the country, Bawdsey Cliff has now been declared a Site of Special Scientific Interest (SSSI) by Natural England.

Harmer stated that the various exposures of the Red Crag group themselves in a horizontal rather than a vertical sequence, with the older deposits occurring towards the south and the newer ones to the north or northeast. He continued that when they are traced from Walton in the southwest to Butley in the northeast, the deposits gradually assume a more boreal and recent character. The earliest indication of the conditions which prevailed during Red Crag times was afforded by a unique and now invisible bed of unstratified grey silt at the base of the Walton cliff section as described by Wood, Jun., in 1864 in which molluscs were found in situ where they had lived and died (Harmer 1910b).

On a visit to the Netherlands Harmer found that there was an area immediately south of the Hoek van Holland which appeared to be experiencing marine and meteorological conditions similar to those which occurred along the East Anglian coast during the Red Crag period, that is, strong onshore prevailing winds but southwesterly in direction rather than easterly—as in the earlier period. He found that a former bay or inlet was nearly choked with shelly sand that was still accumulating on the edges of shoals and in channels formerly occupied by the River Maas; the deposits were often bedded at a high angle with the stratification dipping in different directions as it followed the sinuous windings of the banks. As such drift only now occurs rarely on the shores of East Anglia, its accumulation in such abundance during the Crag period indicates, Harmer believed, that the prevailing winds were predominantly easterly associated with storm tracks aligned at that time further south than they generally occur today. Such evidence of major changes in the wind flow since the Pliocene epoch led Harmer to his pioneering reconstruction of past circulation patterns and study of palaeometeorology.

The identification of the Red Crag by the English palaeontologist, Edward Charlesworth (1813–1893) in 1835 as a separate stage of East Anglian Crag was followed by studies of later geologists, including Harmer, into this Pliocene formation (Reid 1890). Wood, Jun., for example, regarded the shell-beds of Walton-on-the-Naze, comprising strongly-marked southern molluscan fauna, as being the oldest of the Red Crag deposits and grouped the remainder of the formation, in which northern shells occur more or less abundantly, under the names of Suttonian and Butleyan. On the other hand, the renowned British geologist Sir Joseph Prestwich (1816–1896) maintained that all the Red Crag beds were contemporaneous. In view of these differences of opinion, Harmer decided to re-examine the subject to determine, if possible, a more definite conclusion. Accordingly, he spent 5 or 6 years revisiting every part of the Crag region, as well as examining the more important Crag fossil collections with the intention of obtaining evidence to establish the correct classification of these beds.

As a result of his research, Harmer subdivided the Red Crag into three formations on the basis of their fossilized molluscan fauna, namely, Waltonian, Newbournian and Butleyan (Figs. 3.9 and 3.10).

OBLIQUE RED CRAG IN BAWDSEY CLIFF.

The oblique bedded crag appearing above the Talus is nearly twenty feet in thickness.

Fig. 3.9 A formation of the Red Crag in Bawdsey Cliff, Suffolk showing the typical inclination of about 30° to the horizontal of this obliquely bedded deposit (Wood Jun. and Harmer 1872–1874)

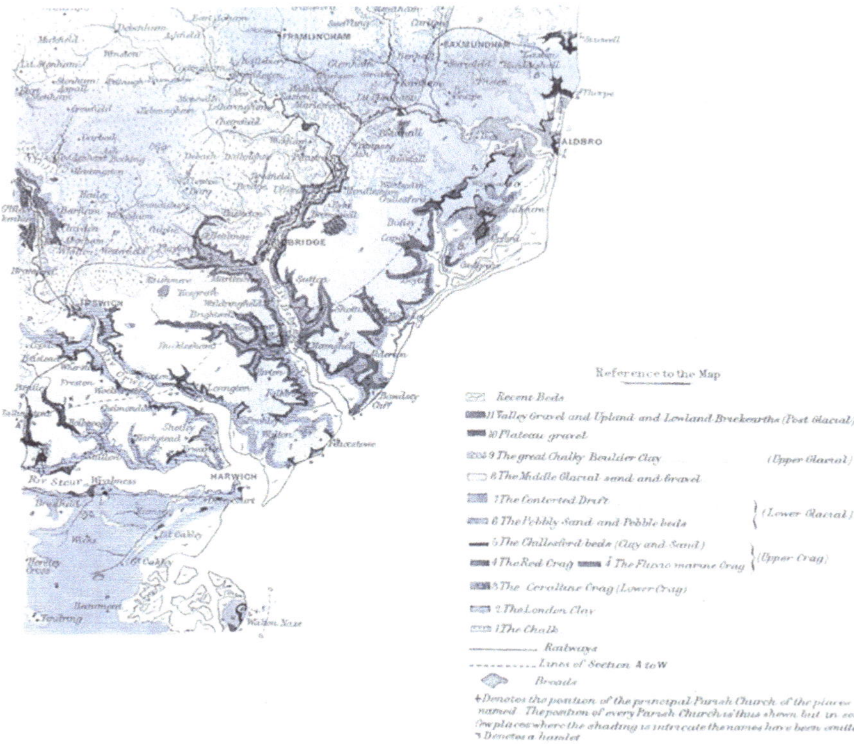

Fig. 3.10 Section of the map made by Wood, Jun. and Harmer showing the upper Tertiary formations of the Crag district, including the Red Crag site in Bawdsey Cliff, Suffolk (Wood Jun. and Harmer 1872–1874)

These formations indicated the progressive retreat northwards and cooling of the Crag Sea, with an ever-increasing proportion of boreal molluscan species. For example, the southernmost exposures of the Red Crag, particularly at Walton-on-the-Naze (appropriately assigned to the Waltonian formation), were considered to represent the oldest Red Crag stage containing the warmest water faunas, including many species that are rare or absent elsewhere in the Red Crag, whilst the Newbournian and Butleyan formations outcropping further north probably indicated cool temperate conditions (Figs. 3.11 and 3.12).

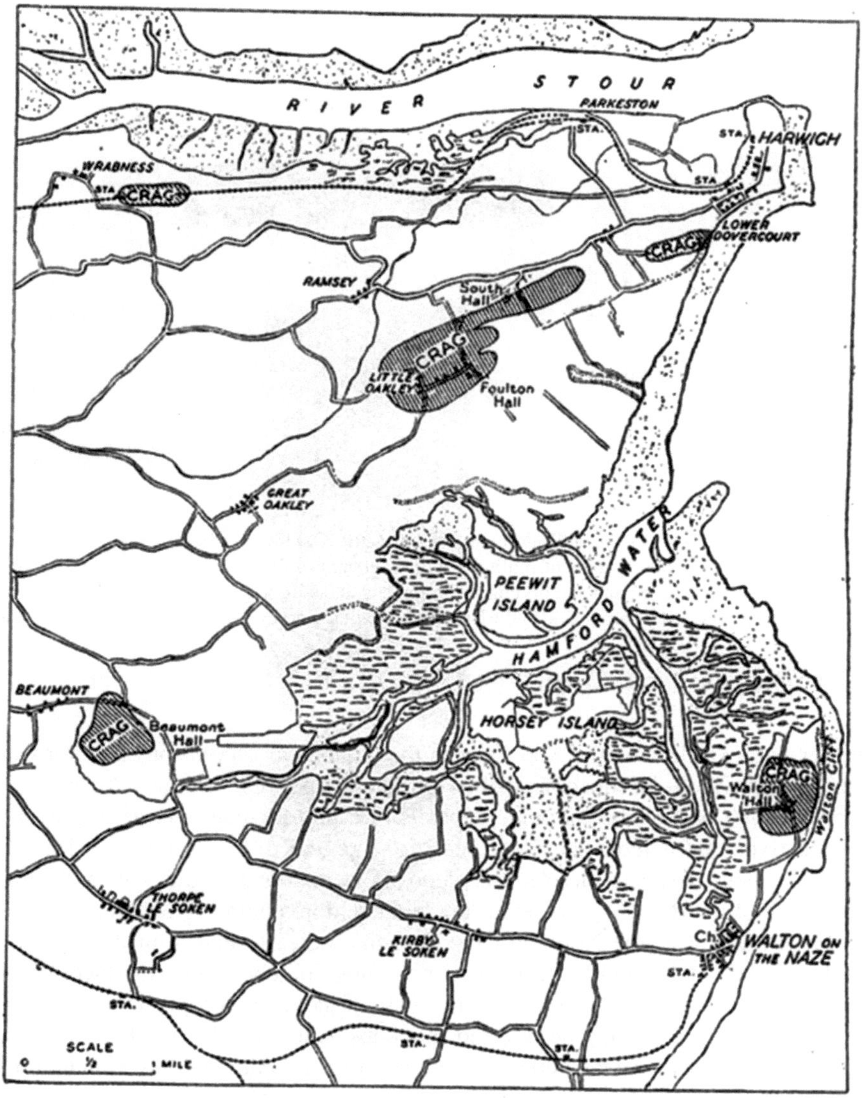

Fig. 3.11 Map of the Waltonian Crag in northeast Essex showing the location of the Pliocene deposits at Little Oakley. Further southeast the cliff sections at Walton-on-the-Naze comprise one of the most accessible and best-known exposures of such deposits in eastern England (Harmer 1900a)

Fig. 3.12 P680273 (1907) Pit of Waltonian Crag at Little Oakley, northeast Essex from which Harmer obtained over 600 species of mollusca within an area of 20 yards square. Harmer's car is visible on the *left middle* ground and his chauffeur appears to be digging out the material for sorting. (CP13/050 Reproduced by permission of the British Geological Survey © NERC. All rights reserved)

3.3.3 The Norwich Crag

Further north (following the Red Crag in the geological succession) the Norwich Crag beds, which lie unconformably on the Upper Cretaceous Chalk, were believed to be part of the estuarine deposits of a proto-Rhine, a large northward-flowing river that entered the Crag Sea in the region of Cromer today (Harmer 1902). In many respects these newer deposits were found to differ considerably from those of the Red Crag. Whereas the latter seldom attain more than about 6 m (20 ft.) in thickness, the Norwich Crag deposits increase rapidly in depth northward being found 45 m (147 ft.) thick, for example, in a boring at Southwold only twelve miles from Aldeburgh. As all the shells discovered at this site were of the shallow-water type, it may be inferred that subsidence of this area had occurred contemporaneously with the deposition of the sediment. Although fewer in number than those of the Red Crag, the character of the Norwich Crag molluscan fauna was found to be generally more or less uniform throughout the whole area covered by the deposits from Aldeburgh to the Bure valley. However a significant regional difference was the presence of the arctic shell, *Astarte borealis* (one of the latest arrivals in the Crag Sea) in northern parts of the area.

The sandy material comprising the Norwich Crag appeared to have originated mainly from a southern source as it contained many pebbles of white quartz, a characteristic feature of Dutch gravels, together with mica, possibly derived from the Devonian schists of the Ardennes. The proto-Rhine with its affluents seems to have been an important factor in the later Pliocene history of East Anglia. This deposit did not present the highly inclined bedding of the Red Crag since it seems to have been laid down under different conditions, probably as the western edge of a large delta of the proto-Rhine. It attained a much greater thickness in the subsoil of Holland where it was deposited in a subsiding area with the subsidence dying out towards the west.

3.3.4 The Chillesford and Cromer Forest Beds, and Weybourne Crag

The next deposits in the geological succession, the Chillesford Beds, overlie the Red and Norwich Crags and appear to be related to the probable course of a proto-Rhine estuary with its exposures of laminated micaceous sand and clay. It is suggested that at two subsequent stages of the late Tertiary, evidence of similar conditions found in the estuarine clays and gravels of the Cromer Forest Bed series containing fossil remains of the Southern Mammoth (*Elephas meridionalis*), also indicate a northward outlet of the proto-Rhine, while in the submerged North Sea deposits the remains of many land animals, such the Northern Mammoth (*Elephas primgenious*) have been dredged from Dogger Bank. Unlike molluscs, these are fossils of extinct species and whilst the mixture of southern and northern forms may indicate changes of climate due to the oncoming cold conditions, there is the

Fig. 3.13 P680260 (1907) Harmer's geological excursion to Norfolk. Cliffs at Runton, Norfolk coast with Cromer Forest Bed deposits at the cliff base. (CP13/050 Reproduced by permission of the British Geological Survey © NERC. All rights reserved)

Fig. 3.14 P680269 (1907) Harmer's geological excursion to Norfolk with Harmer (on *left*), the German geologist, Carl C. Gottsche and the Austrian geologist, Emil Tietze above the Runton Gap, Norfolk coast. (CP13/050 Reproduced by permission of the British Geological Survey © NERC. All rights reserved)

possibility that some of the remains may have been 'derived' that is washed out from material deposited upstream in the proto-Rhine (Figs. 3.13 and 3.14).

The deposition of the Chillesford and Cromer Forest Beds each representing a slight upheaval of the western part of the Pliocene basin, was separated by a period of depression, which led to the formation of the shallow-water marine sands of the Weybourne Crag when the sea re-invaded part of the area now known as Norfolk as far south as Norwich.

The molluscan faunae of the Chillesford and Cromer Forest Beds which were considered to be boreal rather than arctic, were similar to those of the Norwich Weybourne Crag zone, except that they comprised a smaller number of species and contained throughout, in the greatest profusion, a shell, *Tellina balthica*, which up to that time was unknown in the Crag Sea. The relative abrupt and abundant appearance of this species at the Weybourne Crag stage suggested that communication was at that time opened up by a continued progress of the northeasterly subsidence with a sea area, possibly the Baltic, in which this mollusc had established itself (Figs. 3.15, 3.16, 3.17, 3.18, 3.19 and 3.20).

Fig. 3.15 P680268 (1907) Harmer's geological excursion to Norfolk; participants at the base of the cliffs near Weybourne (Harmer sitting on the *right*). Today, the Weybourne cliffs have been designated a Site of Special Scientific Interest (SSSI) with exposures of Weybourne Crag and Cromer Forest Bed deposits overlying the Chalk bedrock. The Forest Bed formation is notable for its Mammalian fossils which have been found along the base of the cliffs and foreshore between Weybourne in Norfolk and Kessingland in Suffolk. Although most of these deposits have now been obscured by coastal defence, further fossils are still occasionally discovered. (CP13/050 Reproduced by permission of the British Geological Survey © NERC. All rights reserved)

Fig. 3.16 P680258 (1907) Harmer's geological excursion to Norfolk. Cliffs at Weybourne, Norfolk comprising Weybourne Crag which overlies the Chalk bedrock; the cliffs are capped by glacial deposits of the Anglian period, a cold stage during the Pleistocene when ice advanced over much of East Anglia. (CP13/050 Reproduced by permission of the British Geological Survey © NERC. All rights reserved)

Fig. 3.17 P680259 (1907) Harmer's geological excursion to Norfolk. Weybourne Crag resting on the Chalk bedrock with stone bed at base, near Sheringham, Norfolk. (CP13/050 Reproduced by permission of the British Geological Survey © NERC. All rights reserved)

Fig. 3.18 P680261 (1907) An exposure of chalky boulder-clay with irregular decalcification in its upper part at Hellesdon, near Norwich. (CP13/050 Reproduced by permission of the British Geological Survey © NERC. All rights reserved)

Fig. 3.19 P680272 (1907) A chalk quarry exposing the Cretaceous bedrock that forms the solid geology of Norfolk; dipping from west to east this formation comprises the top of cliffs at Hunstanton, the base of the cliffs at Weybourne and lies at about 150 m (500 ft.) below sea level at Great Yarmouth. (CP13/050 Reproduced by permission of the British Geological Survey © NERC. All rights reserved)

Fig. 3.20 P680266 (1907) Construction work at Castle Meadow with wooden supports on the slope descending from Norwich Castle; the exposure of chalk and other deposits at this city site would have been of great interest to Harmer. (CP13/050 Reproduced by permission of the British Geological Survey © NERC. All rights reserved)

3.3.5 The Icenian Crag

The Icenian Crag (a term derived from the name of the Iceni, an ancient tribe of East Anglia) includes the Norwich Crag, Chillesford Beds and Weybourne Crag. Whilst the Red Crag is a shore deposit, the formations of the Icenian Crag are believed to have been deposited in shallow waters of the Crag Sea near the estuary of a large northward-flowing river that was probably the forerunner of the Rhine.

The suggestion that there may have been a gradual diminution in the salinity of the Crag Sea during later stages of the Crag period would explain the decline of molluscan fauna in the succeeding Icenian deposits, as compared with its great variety and abundance during the preceding periods. The advance of the ice sheet from Scandinavia in the Icenian period may have gradually obstructed and eventually brought to an end the northern communication of the Crag Sea, while fresh water continued to pour into the sea from the south. Such an influx of fresh water, especially during summer, would have formed a large lake with waters which increasingly became brackish. This might have changed the character of the

Pliocene molluscan fauna of the Crag Sea and eventually may have exterminated a greater part of it. Although communication with the saline waters of the Atlantic Ocean was subsequently re-established by the retreat of the ice sheet in the late Pleistocene and the cutting of the Strait of Dover in postglacial times, the molluscan fauna of East Anglian waters has never regained the richness and variety it possessed during the earlier stages of the Pliocene epoch.

3.4 The Pleistocene Epoch in East Anglia

By compiling and analysing a collective of accounts made by glaciologists in England and Wales, the origin and movement of the ice streams associated with the glacial deposits of East Anglia was determined by Harmer and illustrated by means of a contoured map. This was the first detailed cartographic attempt to show the distribution of erratics and drift in England and Wales in which a system of symbols was employed to indicate the respective sources of these features (Harmer 1928).

Among the geologists who first attempted to resolve the nature of East Anglian drifts, several notable pioneers can be included such as William Buckland (1784–1856), who noticed the presence of Norwegian erratics (1823), R.C. Taylor who called attention to contortions in the coastal sections (1824), C.B. Rose who identified boulders from Mountain Limestone and various Jurassic strata in the 'diluvium' (1835), and Joshua Trimmer (1795–1857), who recognised two types of boulder-clay in the cliff section at Gorleston, the lower identified by Scandinavian blocks and the upper by Oolitic detritus, which he had traced towards the west.

Despite these early 19th century findings, Harmer and others realised that comparatively little had been discovered about the distribution or stratigraphy of the clay, sand and gravel beds which comprised the glacial deposits, as well as the conditions under which they had originated. The only map to show glacial deposits then in existence was one published by Samuel Woodward in 1833 and although the relation of the Crag and 'Diluvium' (an early 19th-century term applied to superficial deposits then believed to be due to a catastrophic biblical deluge) of the coast between Cromer and Great Yarmouth to the underlying chalk had been correctly shown, no attempt had been made up to that time to map the glacial deposits over an extended area, either in East Anglia, or, as far as it was known, in any other part of the world (Figs. 3.21, 3.22, 3.23 and 3.24).

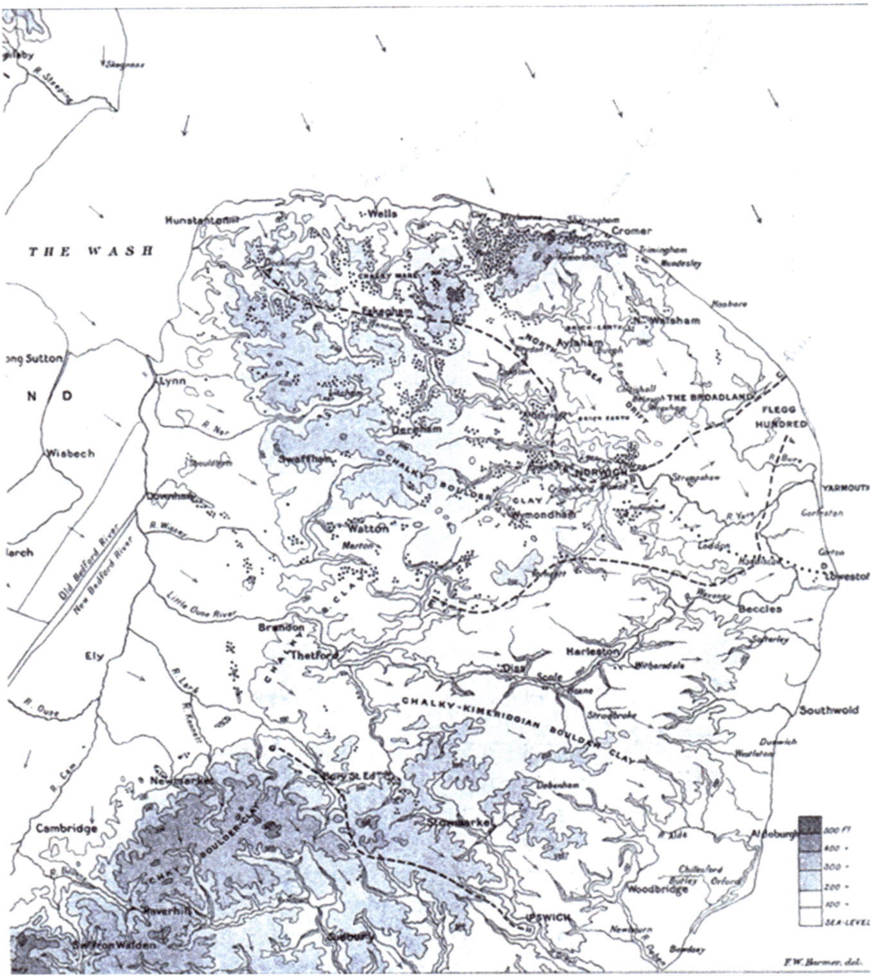

Fig. 3.21 Contour map showing the distribution of the East Anglian drifts. The Cannon-shot gravels are represented by *small dots*. The supposed movement of the ice streams is represented by *arrows* (Harmer 1910a)

Fig. 3.22 P680267 Cannon-shot gravel, Lincolnshire; such deposits also occur in other parts of eastern England, including many sites in Norfolk. The term, 'Cannon-shot' was adopted by Harmer and Wood, Jun. since these gravels occasionally contain symmetrically rounded flints resembling in size and shape cannon-balls that were formerly used in military gun-fire. Harmer suggested that these distinctive flints may probably have been formed during the glacial period in kettle-holes beneath ice sheets. Unlike glacial boulder-clays and sands he noted that they do not form continuous sheets but instead occur in isolated masses, sometimes of considerable thickness and extent, often occupying elevated sites such as at Poringland and Mousehold Heath in Norfolk. (CP13/050 Reproduced by permission of the British Geological Survey © NERC. All rights reserved)

Fig. 3.23 P680265 (1907) This photograph was inscribed, 'striated erratic, Bushbury, near Wolverhampton,' with what appears to be Harmer's signature. (CP13/050 Reproduced by permission of the British Geological Survey © NERC. All rights reserved)

Fig. 3.24 P680278 The Bowder Stone, Borrowdale, Cumbria which is one of the largest erratics found in England; its name probably comes from the Norse god, Baldr (Odin's son). About 9 m (30 ft.) high, this large erratic is estimated to weigh around 2,000 tons and was probably carried to this site by an ice sheet from Scotland. The staircase allowing visitors to climb to the top has been in place since at least 1890. (From Harmer's collection of photographs) (CP13/050 Reproduced by permission of the British Geological Survey © NERC. All rights reserved)

The identification of at least two layers of boulder-clay in Norfolk, separated by interglacial sands, demonstrated that the chronology of the Ice Ages, or the Pleistocene, was more complex than had previously been proposed by the Swiss geologist, Louis Agassiz (1807–1873) in the 1830s. This finding preceded the better known views of the Scottish geologist, James Geikie (1839–1915) who also had suggested that instead of one single advance and retreat, the Pleistocene had been marked by a series of alternating glacial and interglacial stages. Harmer noted that a pit near the Trowse railway junction (one mile south of Norwich) which exposed a glacial deposit of an unstratified whitish clay containing much chalk debris and occasional quartzose erratics, fully provided evidence for this concept (Harmer 1867a, b) (Fig. 3.25).

Fig. 3.25 P680270 (1907) Harmer's geological excursion to Norfolk with a member of the group at a site illustrating distinctive layers of glacial sands and gravels. (CP13/050 Reproduced by permission of the British Geological Survey © NERC. All rights reserved)

Searles V. Wood, Jun. is generally recognised as establishing a systematic study of glaciology in eastern England–East Anglia in particular. He was convinced that many of the problems in understanding the nature of glacial geology could only be resolved by mapping its deposits, an undertaking which, up to that time, had never been attempted in England—let alone in any other country. Wood therefore undertook the monumental task of mapping these deposits on a scale of 1 inch to 1 mile, from the Wash to the Thames valley, and from central England to the Norfolk-Suffolk coast, covering an area of 2,000 square miles.

The following 6–7 years were devoted to examining these deposits with Harmer, who had taken up the challenge set by his colleague with characteristic zeal and industry. Together they undertook this large-scale mapping project to show the distribution of Pliocene and Pleistocene deposits in eastern England, Wood covering the drifts of Suffolk and Essex and Harmer those of Norfolk. By 1868 the work was sufficiently advanced for a geological map to be presented illustrating the distribution of pre-glacial and glacial beds over an area of 2,000 square miles.

Wood and Harmer embodied the results of their study in the memoir, *An Outline of the Geology of the Upper Tertiaries of East Anglia* together with a map (reduced form of the original, with a few corrections) and various sections of the Crag District (Wood Jun. and Harmer 1872–1874), the whole work being included in the 1872–1874 volume of the Palaeontographical Society, *Supplement to the Monograph of the Crag Mollusca, with Descriptions of Shells from the Upper Tertiaries of the East of England* (Fig. 3.6), which remains one of the classic texts in the literature of geology (Wood Sen. 1872–1874).

The definitive distribution of chalky boulder-clay was established by Wood in 1880 (and later Harmer). It was concluded that an ice stream moved from the North Sea across parts of the Lincolnshire Wolds, being deflected southwards by another ice stream from the Vale of York, and then fanned out into East Anglia and central England. A further ice stream of the same glaciation was believed to have entered Norfolk across the present north coast of the county coming into contact with the North Sea Drift (Cromer Till and Norwich Brick-earth).

Harmer believed that the glacial deposits of East Anglia were initially laid down due to the invasion of the region by the western edge of a large ice sheet, such as exists in Greenland today, which originating on the Scandinavian uplands (probably standing at a higher level than today) filled the basin of the North Sea and overflowed across the north European plain. The contorted drift of its *moraine profonde* (a term used at the time for ground moraine) may definitely be traced from the coast at Cromer as far as sites in south Norfolk and Suffolk, such as Sotterley, Withersdale Street, Scole and Bury St Edmunds; it may also have extended even further south, as well as partly spreading over the Fenland.

Harmer continued that at a later stage of the Pleistocene epoch the North Sea ice sheet retreated from East Anglia and although there are no indications that it ever re-advanced, it seems that the edge of the ice remained stationary for a considerable period close to the present coastline, during which time it resulted in the formation of a relatively high and long terminal moraine, the Cromer Ridge. This feature comprises a series of hummocky hills which, in places, reaches over 90 m (300 ft.) high and extends for about 24 km (15 miles) west-southwest to east-northeast across northern Norfolk before ending with the exposed contorted drift in the cliffs between Weybourne and Happisburgh, where it reaches the sea. South of the Cromer Ridge a flat sandy heathland was formed by melt waters during the retreat of the North Sea ice sheet. This outwash plain, a glacial landform not found elsewhere in East Anglia due to its 'newer' drift formation, corresponds to the 'sand plains' of similar morainic origin in northern Germany and Denmark, such as the so-called Heath of Jutland. The terminal moraine in this region traverses the Danish Peninsula from north to south before looping eastwards across the North European Plain (Figs. 3.26, 3.27, 3.28, 3.29, 3.30, 3.31, 3.32, 3.33, 3.34, 3.35 and 3.36).

Fig. 3.26 P680282 (1907) Harmer's geological excursion to Norfolk. An inferred view looking east towards Mundesley showing hummocky hills which probably comprise glacial drift of the extended Cromer Ridge moraine. (CP13/050 Reproduced by permission of the British Geological Survey © NERC. All rights reserved)

Fig. 3.27 P680289 (1907) Harmer's geological excursion to Norfolk. A view from the beach of the cliffs at Cromer. In the background the hummocky landscape of the Cromer Ridge (terminal moraine which can be traced 15 miles WSW across northern Norfolk is truncated by the sea at Cromer; the cliffs at this point rise to about 60 m (200 ft.). (CP13/050 Reproduced by permission of the British Geological Survey © NERC. All rights reserved)

Fig. 3.28 P680290 (1907) Harmer's geological excursion to Norfolk. A view of Cromer pier with its historic Concert Theatre. A substantial buttressed groyne has been built at right angles to the beach to block the eastward movement of sand and shingle by longshore drift; a marked contrast is seen between the shingle on left and the sandy beach on right with several children and a lady apparently digging with small spades for shell fish. (CP13/050 Reproduced by permission of the British Geological Survey © NERC. All rights reserved)

Fig. 3.29 P680291 (1907) Harmer's geological excursion to Norfolk. Looking west along the former Midland & Great Northern Railway line between Mundesley and Weybourne with hummocky drift features in the background near Sheringham. Harmer noted that such knoll-like landforms are also common in other glaciated areas of northern Europe (Harmer 1910a, b). (CP13/050 Reproduced by permission of the British Geological Survey)

Fig. 3.30 P680293 (1907) Harmer's geological excursion to Norfolk. Hummocky drift landforms near Sheringham, Harmer recognised that these rounded hills or knolls were characteristic landforms of a glaciated area (Harmer 1910). (CP13/050 Reproduced by permission of the British Geological Survey)

Figs. 3.31, 3.32 and 3.33 P680283, P680284 and P680285 (1907) Harmer's geological excursion to Norfolk. An inferred view of cliffs in northeast Norfolk between Weybourne and Happisburgh. This section of cliffs mainly comprises glacial sands, gravels and clays which are often twisted and contorted. Land-water draining through the soft strata causes a great deal of slipping and gullying, and notable cliff-falls often occur. The fallen material is eroded by the sea and although the average rate is about 1 m (3 ft.) per year it can occasionally be much greater due to the effect of North Sea storms and wave action. A series of groynes are also visible which have been built to protect the coastline from further erosion. (CP13/050 Reproduced by permission of the British Geological Survey)

◀ **Figs. 3.34, 3.35 and 3.36** P680286, P680287 and P680288 (1907) Harmer's geological excursion to Norfolk. An inferred view of cliffs in northeast Norfolk between Weybourne and Happisburgh showing contorted glacial deposits. Folding takes place by the slumping of material within and on the ice sheet during melting, and by the action of the ice pushing and contorting sediment near its margin. (CP13/050 Reproduced by permission of the British Geological Survey

During the decline and melting of the North Sea ice sheet, valleys in central and western Norfolk were deeply excavated by the action of torrential melt-water streams issuing from the margin of the retreating ice. Harmer would have recognised that the valley in the Ringstead Downs, south of Hunstanton Park, is the result of a former drainage overflow channel that was cut by the melting ice waters during the glacial period. The chalk which forms the bedrock of much of Norfolk is usually covered with drift deposits but in this dry valley white track-ways indicate that the chalk outcrops at the surface.

Later in the Pleistocene and apparently separated from the earlier glaciation by a considerable interval of time, possibly representing one of the mild interglacials proposed by Penck and Brückner, East Anglia was again invaded by ice, this time by a large inland ice sheet from the northwest termed the *Great Eastern Glacier* by Harmer. The ice of this glacier which originated in the uplands of northern England, such as Teesdale, flowed down the Vale of York, and then divided into two ice streams, one branch moving up the valley of the Trent and the other across Lincolnshire and the Fenland, having been reinforced by lateral glaciers descending from the Pennines together with ice which had flowed through the Lincolnshire Wolds or the Humber gap from the North Sea (ice would have lain in considerable thickness off the Lincolnshire coast during that period) (Fig. 3.37).

This gave rise to a widespread deposit, the Chalky boulder-clay, forming the *moraine profonde* of the second glaciation covering a large part of East Anglia as a more or less continuous sheet. It contained a great variety of Jurassic and Cretaceous erratics, especially Kimmeridge Clay, Neocomian sandstones, hard Chalk, and distinctive tabular grey flints from Lincolnshire. This glacier ploughed up and incorporated in its own deposits much of the earlier North Sea Drift, so that the westward extension of the latter is ill-defined.

As a result there are clearly two types of boulder-clay in East Anglia divided by sands and gravels; the most notable feature of the upper one (comprising Chalky boulder-clay with a matrix which varies in different districts according to the various Jurassic rocks over which the ice had passed) contains Pennine and Oolitic detritus and blocks of the characteristic tabular grey flint and hard chalk of the Lincolnshire Wolds. The presence of large quantities of Kimmeridgian detritus could only be explained by its having been transported by ice from the northwest that is in a direction more or less at right angles to the flow of the earlier North Sea Glacier.

A study of the relation of glacial drift in the valleys of the region showed that there had been a great deal of denudation between the depositions of the two boulder-clays. This was of great interest to Harmer in connection with the

Fig. 3.37 P680303 (1907) Harmer's geological excursion to Norfolk. An inferred view of variable glacial deposits in the Norfolk coast showing marked slippage which may have occurred during melting of the ice sheet. (CP13/050 Reproduced by permission of the British Geological Survey © NERC. All rights reserved)

occurrence of inter-glacial periods, since a long interval of time was indicated for such processes to occur which may have corresponded to one of the mild inter-stadials suggested by Penck and Brückner for mainland Europe. Lastly, the origin of the plateau and valley gravels of the region may be ascribed, he claimed, to the torrential waters that were released during later stages of the melting and retreat of the ice.

Harmer summarised the conditions which, he believed, prevailed in England and Wales during glacial times as follows:

> Most glaciologists believe that this country was invaded by ice, on the east from the German Ocean [North Sea], and on the west from the Irish Sea. Crossing the Lincolnshire Wolds, but (as I believe) at certain points only, ice from the North Sea, augmented, I think, by that of an inland glacier from the Vale of York, travelled towards the plain of the lower Witham and the Fenland, whence it overspread a large tract of the country to the east, the south, and the west. To the east it reached the Suffolk coast, to the south nearly to the valley of the Thames, while to the west it filled those of the Welland, the Nene, and the Ouse, overriding also the high land intervening.

Another branch of the northern glacier, keeping to the west of the Lincoln ridge, and reinforced by the North Sea ice, moved towards Doncaster and up the Trent basin to the vicinity of Derby, where it met the Derwent glacier, and thence crept southward along the valley of the Soar into Warwickshire.

In the west, the Cheshire plain was invaded by ice from the Irish Sea which, diverting the glaciers descending from the mountains of North Wales towards the south, carried a large number of Scottish and Lake District erratics into the northern part of the basin of the Lower Severn, heaping them also upon Cannock Chase, and upon the high land near Wolverhampton.

In South Wales, Dr. Strahan and his colleagues have shown that the ice descended in great thickness from the Brecknock Beacons towards the Bristol Channel, reaching the shores of the latter near Swansea, filling the Neath and Taff valleys to overflowing and rising to a great height on the intervening hills.

Evidence has also been found of the invasion of the southern part of this district by ice from the Irish Sea, which is supposed to have travelled up the Bristol Channel from west to east, and to have crossed the Pembrokeshire peninsula from St. David's Head towards Gower and to the neighbourhood of Cardiff: erratics believed to have been derived from the first-named locality having been found nearly 100 miles to the eastward of their probable source.

The depth of St. George's Channel between St. David's Head and Ireland, however, exceeds 50 fathoms, and the natural course of the Irish-Sea glacier, joined by those descending the western slopes of the Welsh mountains, should have been southward along the deep submarine valley opening out to the Atlantic. The distribution of the erratics just mentioned seems therefore to indicate that the volume of ice, approaching the narrowest part of St. George's Channel was too great to enable it wholly to escape in that direction, some of it being forced by lateral pressure to travel eastward up the Bristol Channel.

It seems worth considering whether so important an ice-stream would not have blocked the entrance to the estuary of the Severn, the result being an accumulation of sedentary ice in the valley of that river, derived partly from the glaciers of Central Wales and partly from Atlantic blizzards, which I think, for meteorological reasons, may have prevailed at that epoch. Such blizzards might have filled portions of the Severn basin, then unoccupied by moving ice, with masses of snow that would have eventually consolidated.

This view may possibly throw light on the origin of the great alluvial and lake-like plain of Glastonbury, and of the gorge at Clifton. It may explain also why Arenig boulders [Ordovician age] have been piled up on the Clent Hills, southwest of Birmingham, to a height of nearly 900 feet. It is difficult to understand that this could have occurred, if at that time the Welsh ice could have followed an unobstructed course along low ground towards the Bristol Channel.

The conditions here sketched out, namely, of ice moving upon Central England from the sea in a direction opposed to that of the natural drainage, are precisely those under which Glacial lakes with their accompanying over-flow channels would have naturally originated.

3.5 Pro-glacial Lakes, Cañon-like Valleys and Gorges

Towards the close of the Pleistocene epoch ice sheets over much of Britain had blocked the northward flow of rivers in northern and central England, causing their waters to be ponded back to form extensive pro-glacial lakes such as Lake Lapworth and Lake Humber. At about the same time, rivers normally flowing to the

Wash were also held up by an ice sheet and another large lake was formed, Lake Fenland, which coincided roughly in extent with Fenland today. For a time at least Lake Humber was connected to Lake Fenland by an overflow channel, now marked by the dry Ancaster gap west of Sleaford, and it is probable that the combined glacial waters of these two lakes flowed into the southern North Sea via a Pleistocene antecedent of the present Waveney Valley.

Whilst such pro-glacial lakes are well-referenced in the literature, another example, termed Lake Oxford by Harmer, is not generally included and again illustrates how his geological work has either been forgotten or not known. For instance, without mentioning Harmer, the geographers Wooldridge and Morgan (1937) state:

> It has been suggested that Goring Gap was initiated by the overflow of a suppositious "Lake Oxford," thus uniting the Upper and Lower Thames drainage. The hypothesis remains worthy of investigation but presents many difficulties.

Although these authors refer to 'F.W. Harmer, Q.J.G.S. 1907 (vol. 63)' in their 'Bibliographic Notes', they do not mention the title of his key paper, 'The origin of certain cañon-like valleys' as published in the *Quarterly Journal of the Geological Society*, **63**, 1907, 470–513.

Following reference to some of these pro-glacial lakes and cañon-like valleys, Harmer discussed another less acknowledged such feature which he termed, Lake Oxford. Glacial drift pebbles found in Gloucestershire over 32 km (20 mi) from Oxford are believed to have been deposited close to the former shore-line of this pro-glacial lake (Richardson 1910). Such deposits apparently provide evidence of the extent of this glacial water body over central-southern England during the close of the Pleistocene period due to the damming up of its waters by ice filling its former outlet valleys.

In connection with the development of the Great Ouse basin, Harmer, following Prestwich, suggested that prior to the glacial period, the drainage of the Oxford region had been directed northeast towards the Wash via Bedford. However, later, as a result of the advance of North Sea ice, this drainage was ponded back, forming an extensive lake (Lake Oxford) which ultimately found an outlet over the Chalk escarpment of the Chiltern Hills at Goring, diverting a large amount of drainage into the synclinal valley of the original Thames. This was of interest to Harmer due to its bearing on the present oversize valley of the Great Ouse and the deep excavation of the Fenland area both of which, he believed, had been caused by the great volume of water carried down by the river to the Wash during pre-glacial times, that is before the drainage of the upper part of its basin had been lost by diversion into the Thames (Harmer 1907).

As the present Great Ouse in North Buckinghamshire appears to be too small for its wide valley it can be described as a misfit stream with the implication that its discharge today has in some way been reduced from that of former times. In 1947 the author witnessed the rapid transformation of this small misfit stream into a much larger river when, due to exceptional flooding, the valley was completely filled. This occurred in March that year when a sudden thaw of heavy winter snow and ice caused widespread river flooding in England. This extreme situation may provide an analogue for the pre-glacial Great Ouse when it was powerful enough to carve out its present over-size valley before its upper catchment area was blocked by an advancing ice sheet from the north.

When Harmer discussed the glacial origin of some gorge-like valleys in England, he suggested that such cases must be typical and not anomalous. Harmer continued that in all glaciated regions, whether in Britain or abroad, the invasion of any region by an ice sheet, especially where its movement was upstream, would obstruct the natural drainage, producing lakes in which levels would have risen until an outlet for the water was established in some new direction. In this key paper, Harmer presented the hypothesis that certain gorges in England might have originated as spillways due to the blocking of the previously normal line of drainage by an ice sheet (Harmer 1907).

Harmer's hypothesis was later referred to by Leonard Wills when he suggested that the Upper Severn drainage was added to that of the Lower Severn due to the impounding of a glacial lake in Shropshire and Cheshire which rose until it flowed over the former watershed between the Shropshire Plain and the headwaters of a tributary of the Worcestershire Stour (Wills 1924).

In 1947, Sherlock also made reference to Harmer's hypothesis, when, in discussing gaps in the Chiltern Hills, he suggested that an ice-sheet advancing from the north during the Pleistocene epoch impinged on the escarpment but was not able to override it in the Wendover-Tring section. With the waning of the ice, massive floods of water were released forming a lake between the Chiltern Hills and the retreating ice. The waters of this lake rose until their level reached cols in the escarpment where Wendover and Tring stand today. Here they overflowed into valleys with torrents of great erosive power on the dip-slope of the Chiltern plateau cutting deep steep-sided gorges very rapidly into the soft chalk—fully vindicating the hypothesis suggested by Harmer 40 years earlier (Sherlock 1947).

Although the geological history of the Thames remains the subject of conflicting hypotheses, in 1907 Harmer drew attention to an interesting parallel between the basin of the Middle Thames around Oxford and the Vale of Pickering in Yorkshire (Harmer 1907). The latter, as is well known from Kendall's 1902 paper, was occupied by a glacial lake which was due to the Derwent River having had its outlet to the sea south of Scarborough closed by a dam of ice; as a result the water in this lake rose until it overflowed at a gap near Malton; the river thus formed cut a gorge through which the drainage from the Vale of Pickering now flows southwestward into the Yorkshire Ouse, and reaches the sea through the Humber.

According to Harmer the Upper Thames originally discharged northeastward through the Fens into the Wash; however when this outlet was blocked by ice, the waters of the Upper Thames collected as a pro-glacial lake (Lake Oxford) which were discharged by several overflow channels cut through the Chiltern Hills; eventually, as the lake-level fell, the discharge was only maintained through the Goring Gap at the southwestern end of the Chiltern Hills.

In suggesting possible pre-glacial drainage patterns for eastern England, Harmer anticipated current research regarding the reconstruction of pre-glacial river systems which has shown that the region was drained by a major easterly flowing river (possibly the Pleistocene 'Bytham River') which joined the North Sea near Lowestoft (Bricker et al. 2012). However, recent research has shown that the 'Bytham River' could not have existed in the form earlier suggested and that further examination of ancient drainage patterns in eastern England needs to be undertaken (Gibbard et al. 2013). With his studies of variations in river flow before and during the glacial epoch, Harmer would certainly have been interested to know that the subject of variations in drainage patterns is currently being investigated by British geologists.

Harmer also suggested that it would be interesting to investigate whether the origin of two notable gaps in the Jurassic escarpment at Lincoln and Ancaster, which connect in an apparently artificial manner, the basins of the Trent and the Lower Witham, could be explained by applying the theory he had proposed concerning major changes in river drainage due to ice-sheet action. Writing with reference to Harmer's theory, Jukes-Browne stated:

> The Honington [Ancaster] gap is, in fact, one end of a transverse valley which cuts through the Jurassic escarpment, and the tract of gravel west of Ancaster is now the watershed between the Honington Beck flowing westward and a nameless beck flowing eastward to Sleaford. Like most other transverse valleys, it probably owes its origin to a stream which ran eastward across the Jurassic area before the escarpment was developed, and in that case it would be of pre-glacial age. Mr. F.W. Harmer (1907), however, has recently suggested that both the Honington [Ancaster] and the Lincoln gaps were formed during the Glacial epoch by the overflow of a large lake [Lake Humber] the formation of such a lake being due to the advance of a massive ice-sheet from the north which dammed up the outlet of the Trent and other rivers.
>
> Mr. Harmer thinks that the lake so formed first found an outlet at Lincoln, and when this was blocked by the further advance of the ice a second one was formed along the Honington [Ancaster] gap. The exact way, however, in which these two gaps were formed has nothing to do with the inference from the disposition of the gravels that in early post-Glacial time the Witham ran through the Honington [Ancaster] gap and not through that at Lincoln. .. In the same way there can be little doubt that at the time when the Witham flowed through the Honington and Ancaster valleys the Trent flowed through the Lincoln gap (Jukes-Browne 1910).

3.6 The Pre-glacial Course of the Thames

In early and middle Pleistocene times (about a million years ago) the Oxford basin was drained by two rivers, the proto-Thames and its tributary, the Bytham with the proto-Thames flowing northeastwards. During the Anglian cold stage of the last glaciation, ice sheets to the north were melting and a pro-glacial lake was formed to the north of Goring (termed Lake Oxford by Harmer); later its overflow channel eroded a gorge through the Chalk escarpment south of Oxford to create the Goring Gap. The proto-Thames now flowed south through this gorge and then east towards what is now London, where it became a tributary of the Rhine; Britain at this time was still connected to mainland Europe. Harmer continued:

> The features of that part of the Jurassic plain which lies east and west of Oxford and those of its outlet through the Chalk-escarpment, the well-known Goring gap, are so nearly identical with the cases before discussed [that is, lake-like basins drained by gorges excavated transversely across adjacent hill ranges] as to suggest primâ facia that they originated in a similar way.
>
> The basin in question [the Oxford plain] is bounded on the north-east, between Buckingham and Leighton Buzzard, by a transverse ridge of comparatively high land, from 400 to 500 feet above sea level—partly composed, however, of Glacial Drift, that is the terminal moraine of the Ouse branch of the great Eastern Glacier. At present, this ridge marks the division between the drainage-system of the Upper Thames and that of the Fenland, but (as before suggested) the watershed may have lain farther to the south-west in pre-Glacial times, the Oxford plain then standing at a somewhat higher level than it now does.
>
> Some deep borings at Hitchin and at Newport (Essex), revealing Glacial drift at a depth of 68 feet in the one case, and 140 feet in the other, below Ordnance-datum, point to the existence in pre-Glacial times of valleys descending obsequently from the Chalk-escarpment.
>
> In another boring 2 miles south-west of Sandy, in Bedfordshire, on low ground near the Greensand-escarpment, Boulder-Clay 104 feet thick was found overlying the Oxford Clay, about 100 feet below the level of the Bedfordshire plain and but little above that of the sea.
>
> Such cases seem to indicate pre-Glacial valleys, subsidiary to that of an important river flowing in a north-easterly direction towards the Wash, at a level considerably lower than the River Ouse at present.
>
> Further to the north-east borings at Boston, Fossdyke and Long Sutton in which Boulder-Clay was shown to extend from 100 to 160 feet below sea-level, may represent the seaward extension of such a drainage-system, excavated at a time when England stood higher than it does now.
>
> Approaching more nearly the Oxford region, we find a boring near Stony Stratford [North Buckinghamshire], in the upper part of the Ouse basin, with 112 feet of Drift, resting upon the sub-Glacial surface at about 115 feet above Ordnance-datum. The present level of the nearest part of the Oxford plain is 100 feet higher, suggesting that the drainage of the latter may have been originally towards the Fenland through this Drift-filled valley and some channel now hidden by Glacial deposits, which are in that region of considerable thickness.
>
> At Buckingham, for example, 6½ miles south-west of Stony Stratford, a well was carried for nearly 70 feet through glacial gravels to a level of about 220 feet O.D., without piercing them; and another boring in the same locality showed 78 feet of Drift.
>
> Assuming that the borings at Buckingham and Stony Stratford represent portions of the same pre-Glacial valley, it is improbable that the former coincides with its maximum depth near that place, as the longitudinal rivers of the Jurassic plain have no such fall as 100 feet in 6 or 7 miles; under any circumstances, the valley could hardly have terminated abruptly near Buckingham.
>
> If, moreover, the Glacial deposits could be removed, it would be found that the Oxford plain was formerly connected with the basin of the Ouse by a valley of a more important character, wider as well as deeper, than that which still exists. Such a valley seems to have no meaning, and its origin is not easy to explain, unless we regard it as excavated by a river flowing from the south-west; in fact, by the primeval Thames.
>
> The late Sir Joseph Prestwich, indeed, maintained that the Thames (or Isis), originated in pre-Glacial times, ran towards the Wash, having been diverted to the south during the Glacial Period by the excavation of the gorge at Goring.

> A line of low hills, occasionally, as near Oxford, capped by gravel, extend from Purton by Faringdon and Wytham to Headington and Beckley, and thence to Brill; they seem to represent isolated portions of a former Oolitic escarpment, principally Corallian. Such an escarpment, then perhaps unbroken, might possibly have formed the southern slope of the valley in which the pre-Glacial river ran.

3.7 Amstelian Stage

> The name Amstelian, suggested to me as an appropriate one by Dr Lorié, is taken from that of the river Amstel, upon which stands the city of Amsterdam, where these beds have their greatest known development (Harmer 1896a, b, c).

Amstelian, the term given by Harmer to a now abandoned geological stage, was dated about 2½ million years ago. It was geologically important because with its introduction by Harmer a severe cooling became established for the first time in Holland by means of marine faunal assemblages. This decrease in temperature was apparent from the fossil molluscs that had been found by the Dutch geologist, Jan Lorié in a series of deep boreholes in Holland, including one of 156 m (513 ft.) below sea-level at Utrecht and another of 190 m (625 ft.) at Amsterdam (Harmer 1910b).

As Harmer was recognised as the authority on Pliocene fossil molluscs in England, Lorié sent him some relevant papers on the subject regarding the strata found in these deep borings. Harmer noticed from Lorié's descriptions of the fossil content of the strata, the presence of boreal-arctic and extinct molluscan species which, although he (Harmer) knew had occurred in England, had not, up to then, been found in Holland. Harmer therefore proposed to introduce a new geological stage and, on the suggestion of Lorié, this became known as the Amstelian.

When shown in section, these borings revealed that the Pliocene beds underlying Holland not only attained a thickness of about 150 m (500 ft.), but that they had been persistently and gradually depressed until, at the farthest point north to which the borings were carried out, they were found to reach a depth of more than 300 m (1,000 ft.) below their original position. This northerly subsidence seems to have been coincident with an elevated earth-movement to the south by which the Lenham-Diestien beds gained a maximum height of more than 180 m (600 ft.) above sea-level; the line of greatest disturbance by the earth-movement running at right angles to the strike from the Straight of Dover to Amsterdam. A comparison of the mollusca found at different levels in these Dutch borings pointed to the conclusion that whilst those found in the lower part corresponded more or less with the characteristic fossils of the Diestien, Casterlien and Scaldisien deposits of Belgium, those of the upper portion were generally of a newer character. For the latter, Harmer proposed (as mentioned above) the term Amstelian.

Although initially the Amstelian stage was considered to be of Pliocene age it was later decided that a deposit with marine fauna containing such boreal-arctic mollusca should be included in the Pleistocene. Since this ambiguity proved to be confusing for subsequent investigations it was decided that the term Amstelian should be omitted from the geological succession.

Unfortunately, further studies of the Pliocene and Pleistocene deposits of eastern England by Wood, Jun. were arrested by his failing health. Nevertheless, he was able to show by means of an original map published in 1880, 4 years before his death, the extent of the area covered by Chalky Boulder Clay in East Anglia. This notable map was also used by Harmer to illustrate his own paper about the glacial deposits of Norfolk and Suffolk published in 1910.

Sadly, the loss in 1884 of his friend and scientific associate was a blow which adversely affected Harmer's enthusiasm for geological research and, for a time, he devoted himself more to municipal duties and local politics. However, although he had been deeply upset by the loss of his friend, about 10 years later when, at the age of 60, he might well have felt entitled to enjoy his retirement, Harmer resumed an intensive study of the Pliocene and Pleistocene deposits of East Anglia and mainland Europe on discovering that his views had found a sympathetic hearing among a younger generation of geologists.

3.8 Renewed Geological Research

In 1902 Harmer was requested by the Geologists' Association to prepare a résumé of his views on the later Tertiary deposits of eastern England; the resulting paper, based on observations made on a long excursion from 26 July to 4 August that year, inaugurated a new chapter in the geology of the region.

Also during the early 1900s, Harmer took on the task of addressing a general oversight in the search for evidence of glaciation in southern England by traversing the region from the Humber to the Thames and from the East Coast to the Welsh Border in his new motor car (Fig. 3.38).

> Most of the area over which the glacial ice is supposed to have passed has been examined personally by myself with more or less care with the aid of a motor car. During several of the excursions I had the advantage of being accompanied by the late J. Lomas, Dr R.H. Rastall, Mr G.H. Slater, and Professor P.G.H. Boswell, and on one occasion by Professor P.F. Kendall.

Harmer recognised that these later investigations were 'rendered possible to an old man by the fortunate invention of the motor car, which literally gave me a new lease of geological life for field work'. In the early 1900s, people who had the means and need replaced their horse-drawn carriages with chauffer-driven motor cars.

Harmer acquired a personal and intimate familiarity with glacial deposits which was probably not attained by any other geologist of his time. Sadly, the opportunity to examine the geological sites which had been available both in his earlier studies and those during the latter part of his life would perhaps never recur due to their ever increasing loss by man's activities in the following years. Probably to counteract this effect he organised the already mentioned field excursion to Norfolk in 1907 for many of his geological contemporaries.

Fig. 3.38 P680274 Harmer being driven in his motor car by his chauffeur. Harmer wrote, 'my later investigations, rendered possible to an old man by the fortunate invention of the motor car, which literally gave me a new lease of geological life for field work (Harmer 1910). (From Harmer's collection of photographs) (CP13/050 Reproduced by permission of the British Geological Survey © NERC. All rights reserved)

In 1910, in recognition of his renown as one of the leading authorities on the geology of eastern England, Harmer was asked to contribute to the Jubilee Volume of the Geologists' Association. In the two papers that followed Harmer took the opportunity to carry out a comprehensive review of his research on the Pliocene and Pleistocene epochs in East Anglia which he had undertaken during many previous years.

The Frederic William Harmer Collection of Crag Mollusca (including material figured in his Palaeontological Society Monographs) is included in the reference material at the Sedgwick Museum of Earth Sciences, Cambridge.

Harmer believed that the sculpturing of East Anglian landforms was mainly due to the particular conditions associated with the melting of the ice sheet, conditions which today have wholly ceased to exist. Writing in 1910, he thought that comparatively little erosion or deposition had occurred in this region since the final disappearance of the ice and continued that the Cromer Ridge moraine may not greatly differ in appearance from that which it presented when the ice left. He stated that the boulder-clay plateaus of the higher parts of Norfolk and Suffolk are as level as if some immense steam roller had recently passed over them and that no great sheets of gravel are accumulating at present along our inland valleys, nor is any river erosion now taking place; the valleys indeed are in fact being gradually filled up with silt and alluvium. He concluded:

> An era of geological rest has settled down on the fertile lands of East Anglia, and we now enjoy in peace the fruits of the glacial disturbance and turmoil of the past.

However, Harmer would have been concerned with the present 'disturbance and turmoil' of his once unspoilt surrounding countryside due, not this time to natural

Fig. 3.39 P680292 (1907) Harmer's geological excursion to Norfolk. A Norfolk village similar to Cringleford during the early 20th century. (CP13/050 Reproduced by permission of the British Geological Survey © NERC. All rights reserved)

Fig. 3.40 P680281 (1907) Harmer's geological excursion to Norfolk. A village in Norfolk. (CP13/050 Reproduced by permission of the British Geological Survey © NERC. All rights reserved)

geological processes, but to the impact of man with his burgeoning building and infrastructure developments—none more so than in his home village of Cringleford where hundreds of new houses are being built on former agricultural land.

Also, although still standing today, Harmer's imposing Victorian house has been divided into two dwellings and his beautiful and extensive gardens sloping down to the river Yare are now covered with a maze of roads and houses (Figs. 3.39 and 3.40).

In 2013 a paper by the author about the scientific work of Harmer was published, appropriately, in a journal devoted to articles about the geology of his home region, East Anglia (Kington 2013).

References

Bricker, S.; Lee, J.; Banks, V.; Morigi, A.; Garcia-Bajo, M., 2012: "East Anglia's Buried Channels", in: *Geoscientist*, May: 15–19.
British Geological Survey, 2013.
Gibbard, P.L.; Turner, C.; West, R.G., 2013: "The Bytham River Reconsidered". *Quaternary International*, 292: 15–32.
Harmer, F.W., 1867a: "On the Existence of a Third Boulder-Clay in Norfolk" in: *Quarterly Journal of the Geological Society* 23: 87–90.
Harmer, F.W., 1867b: "Drift of the Eastern Counties", in: *Geological Magazine*, 4.
Harmer, F.W., 1896a: "The Pliocene Deposits of Holland and Their Relation to the English and Belgian Crags, with a Suggestion for the Establishment of a New Zone, 'Amstelian' and Some Remarks on the Geographical Conditions of the Pliocene Epoch in Northern Europe", in: *Quarterly Journal of the Geological Society*, 52: 748–782.
Harmer, F.W., 1896b: "The Southern Character of the Molluscan Fauna of the Coralline Crag". *Geological Magazine*, 33.
Harmer, F.W., ca 1896c: Draft Paper, 'On Middle Glacial Sands, Coralline Crag and note from Professor Kendall' (BGS).
Harmer, F W., 1900a: On a Proposed New Classification for the Pliocene Deposits of the East of England. Report of the British Association for the Advancement of Science, Transactions of Section C (for 1899): 751–753.
Harmer, F.W., 1902: "A Sketch of the Later Tertiary History of East Anglia" in: *Proceedings of the Geologists' Association*, 17: 416–451.
Harmer, S.F., ca 1904: Draft Paper, 'Tertiary Fossils of the Lower Rhine' (BGS).
Harmer, F.W., 1907: "On the Origin of Certain Cañon-Like Valleys Associated with Lake-Like Areas of Depression", in: *Quarterly Journal of the Geological Society*, 63: 470–514.
Harmer, F.W., 1910a: "The Glacial Deposits of Norfolk and Suffolk", in: *Transactions of the Norfolk and Norwich Naturalists' Society*, 9: 108–133.
Harmer, F.W., 1910b: "The Pliocene Deposits of the Eastern Counties of England", in: Monckton, H.W.; Herries, R.S. (Eds.): *Geology in the Field: The Jubilee Volume of the Geologists' Association (1858–1908) Part I* (London: Edward Stanford): 86–102.
Harmer, F.W., 1928 (posthumously): "The Distribution of Erratics and Drift", in: *Proceedings of the Yorkshire Geological Society*, 21: 79–150.
Kington, J.A., 2013: "Frederic William Harmer: Geologist and Pioneer Palaeometeorologist", in: *Bulletin of the Geological Society of Norfolk*, 63: 13–42.
Challinor, J., 1971: *The History of British Geology* (Newton Abbot: David & Charles).
Dale, S., 1704–1705: "A Letter from Mr Samuel Dale to Mr Edward Lhwyd, Keeper of the Ashmolean Repository in Oxford, Concerning Harwich Cliff, and the Fossil Shells There, Braintree, February 1703", in: *Philosophical Transactions of the Royal Society*, 24: 1568–1578.
Reid, C., 1890: *The Pliocene Deposits of Britain* (London: H.M.S.O).
Richardson, L., 1910: "The Neozoic Rocks of Gloucestershire and Somerset", in: Monckton, H.W.; Herries, R.S. (Eds.): *Geology in the Field: The Jubilee Volume of the Geologists' Association (1858–1908) Volume I* (London: Edward Stanford): 329–364.
Sherlock, R.L., 1947: *British Regional Geology: London and Thames Valley*, London: (H.M.S.O)

West, R.G., 1957: "The Pleistocene Epoch in East Anglia", in: *Journal of Glaciology*, 3: 211–216.

Wills, L.J., 1924: "The Development of the Severn Valley in the Neighbourhood of Ironbridge and Bridgnorth", in: *Quarterly Journal of the Geological Society* , 80: 274–314.

Wood, S.V., Sr., 1872–1874: *Supplement to the Monograph of the Crag Mollusca, Volume III, Univalves and Bivalves* (London: Palaeontographical Society).

Wood, S.V. Jun.; Harmer, F.W., 1872–1874: "An Introductory Outline of the Geology of the Same District, and Map", in: Wood, S.V. Sr. (Ed.): *Supplement to the Monograph of the Crag Mollusca, Volume III, Univalves and Bivalves*, (London: Palaeontographical Society).

Wooldridge, S.W.; Morgan, R.S., 1937: *The Physical Basis of Geography: An Outline of Geomorphology* (London: Longmans, Green and Co.).

Chapter 4
Palaeometeorology

Abstract Review of Harmer's pioneer palaeometeorological research based on the reconstruction of past circulation patterns including reference to climatic changes during the Tertiary and Quaternary periods; formation of the East Anglian Crag deposits; differences in meteorological conditions during the Pliocene epoch compared with today; blocking anticyclones; travelling depressions; changes in storm tracks; variations in the prevailing winds over East Anglia and southern North Sea; comparison of charts by Harmer and Lamb for winter and summer seasons during the Ice Age; possible changes in the distribution of land and sea; comparison of past and present temperature gradients; increased storminess and rainfall during the Pleistocene in the Mediterranean region.

Keywords Palaeometeorology · Tertiary and Quaternary periods · Past circulation patterns · Synoptic analysis · Formation of the East Anglian Crag · Molluscan fauna · Anticyclones · Depressions · North Atlantic Drift

4.1 Introduction

The distribution of the earth's landmasses and oceans into a geographical pattern that would allow the development of climatic refrigeration occurred in the Tertiary period and progressive cooling, which eventually led to the Pleistocene glaciations, set in during the middle and late Miocene epoch; by the Pliocene epoch a cool temperate climate prevailed in the British Isles. Major changes in the oceanic circulation of the North Atlantic and the northward drift of the European landmass were two further effects that contributed to the progressive cooling in that region.

Through his studies of the Tertiary and Quaternary periods Harmer became interested in palaeometeorology and pioneered the reconstruction of past circulation patterns, a valuable tool in meteorological analysis providing a synoptic or 'bird's eye' view of atmospheric behaviour. He was convinced that the major changes of climate that occurred during the Pliocene and Pleistocene epochs were

J. A. Kington, *Frederic W. Harmer: A Scientific Biography*, SpringerBriefs in Environment, Security, Development and Peace 19, DOI: 10.1007/978-3-319-07704-8_4,

linked to long-term variations in the general circulation which could be investigated in a novel and useful way by the reconstruction of past circulation patterns.

As with his innovative ideas in the business world, Harmer's application of palaeometeorological methods to geology again illustrates his pioneering endeavours—this time in science.

He questioned that although in speculations about the causes of anomalous past climates, especially those of the Pleistocene epoch, geologists had recognised the crucial role played by ocean currents, as well as the influence of winds upon the latter, they had seldom questioned how climatic changes may have also been driven by variations in the wind flow itself, that is, by changes in past circulation patterns.

Harmer believed that by combining findings of geology and meteorology, past circulation patterns could be determined which would provide an effective means for solving certain geological problems. Anomalous past climates, he believed, could only have occurred when atmospheric situations prevailed which were different from those normally experienced in the present period.

His attention had been drawn to palaeometeorology by chance when endeavouring to solve an outstanding geological problem, namely, the meteorological conditions under which the Crag deposits of East Anglia had been laid down. Some of these beds originated as littoral accumulations during the Pliocene on the western margin of the proto-North Sea, termed the Crag Sea. Comprising a mixture of sand and molluscan shells these deposits are found today in a more or less continuous line of land-locked strata in the cliffs of the East Anglian coast; 2 million years ago the shore-line of the Crag Sea extended further west than that of the present North Sea.

4.2 The Red Crag

Harmer was particularly interested in the Red Crag formation, a shore-line deposit comprising a succession of beds varying in thickness up to about 6 m (20 ft) with laminae of sands and shells inclined at a high degree to the horizontal. He had discovered that the molluscan shells of these deposits are seldom found today on the shores of East Anglia and that one may walk for miles along the beaches of the region without finding more than a chance specimen. However, this is not due to any absence of such molluscan life in the adjoining sea which Harmer, assisted by his son, Sidney, had ascertained by dredging.

Accordingly, he inferred that the meteorological conditions under which the Red Crag was deposited must have differed from those prevailing today and suggested that easterly rather than westerly winds may have prevailed over East Anglia and the southern North Sea in the earlier period. Investigating the matter further, he ascertained, during a visit to Holland, that the beaches there were, in places, plentifully strewn with molluscan shells, probably comparable in amount and extent with those that had accumulated in East Anglia during the late Pliocene.

Harmer realised that in the present period Atlantic depressions which approach Western Europe mostly travel east over the North Sea between eastern England and Holland. As a result, strong westerly winds and rough seas which generally prevail in the southern sector of these low pressure systems would frequently affect the Dutch coast.

The climate of the Pliocene appears to have undergone gradual refrigeration, as indicated by changes in the molluscan fauna of the Crag Sea. Harmer noted that although species were characteristically southern during the early Pliocene (resembling those of the Mediterranean in the present period) the sea was gradually invaded by boreal mollusca which eventually completely supplanted the southern fauna in the Red Crag as the climate became cooler on the approach of the Pleistocene glaciation. He found, for instance, that species now only found north of the Arctic Circle, became established in the Crag Sea as far south as 52° N.

Harmer suggested that the climate of regions lying to the north of the British Isles during the late Pliocene probably became considerably colder than today as ice sheets and snowfields were established over Scandinavia, leading to the formation of permanent anticyclones. He proposed that these high pressure patterns which later extended across the British Isles and northern Europe during the Pleistocene may be compared, but on a different time-scale, with the semi-permanent anticyclones which form in winter during the present period over the continental regions of northern Eurasia and North America. Under such so-called blocking high conditions travelling depressions approaching the British Isles from the North Atlantic take a more southerly track than usual with stormy weather and strong easterly winds frequently occurring over eastern England and the southern North Sea. Harmer described how such conditions in the Pliocene epoch might have affected northeast Essex where he had spent several years carrying out field studies of the Red Crag.

> It is not difficult to understand, when standing on the cliff at Dovercourt, looking out to sea over the bay formed by the projecting headlands of Walton-on-the-Naze on one side, and of Felixstowe on the other, that it would need only the prevalence of strong easterly winds to reproduce there the conditions of the Red Crag period. Were the direction of the winter gales which now drive vast quantities of sand and dead shells on to the shores of Holland, turned towards Essex, the mud flats of Dovercourt Bay would soon become silted up with shelly sand, as was that of the Waltonian region at a former period (Harmer 1902).

4.3 Weather and Climate in the Pliocene and Pleistocene Epochs

In 1901, Harmer published a landmark article on palaeometeorology in the *Quarterly Journal of the Geological Society*, 'The influence of the winds upon climate during the Pleistocene Epoch: A palaeometeorological explanation of some geological problems'. This paper (read on 8 May 1901 at the *Proceedings of the Geological Society*) was one of the few dealing with palaeometeorology that had been published in a geological journal up to that time, which, in Harmer's case, indicated how certain problems in geology could possibly be resolved by means of meteorological methods. He believed that advances made in meteorology during the late 19th century in synoptic weather analysis could be adapted to resolve such questions (Harmer 1901).

> Winds are an important factor in determining the distribution of climatic zones. Deviations of the isotherms from the normal are generally connected with the direction of the prevailing winds. The influence of marine currents is indirect rather than direct. Changes of wind cause marked and sudden changes in the weather, though the general direction of ocean currents remains the same. Permanent alterations in climate during past epochs would have equally resulted from permanent changes in the wind. Anomalous weather is due to some unusual arrangement of high and low pressure areas. Former cases of anomalous climate can only have occurred when the meteorological conditions were favourable.
>
> Continental areas tend to be cyclonic in summer and anticyclonic in winter, while the reverse is broadly true of the oceans. During the Glacial Period ice-covered areas would have remained more or less anticyclonic throughout the year, while low pressure areas must have prevailed in regions to the south of them and over the adjoining oceans. This would have altered the prevalent direction of the winds and the distribution of rainfall; thus the anticyclone of the European ice-sheet may have caused cyclonic storms to pass farther south than at present, bringing oceanic winds over the Sahara, which formerly enjoyed a humid climate. Dead shells are rarely found now on the eastern shores of Norfolk and Suffolk, though they are driven on to the Dutch coast by westerly gales. Shell-débris in the Upper Crag-beds of East Anglia shows that easterly gales were common at that period. This may have been due to the altered path of cyclones, caused by the glacial conditions which were becoming established in regions to the north of Great Britain (Harmer 1901).

To highlight his approach, Harmer published selected charts illustrating his reconstruction of circulation patterns for the winter and summer seasons during the maximum glaciation of Europe about 18,000 years ago when northern and central parts of the British Isles were covered by a thick and extensive layer of ice (Kington 2010). By means of these charts he inferred that travelling depressions, which at present generally move west-east over the Atlantic towards Europe, must have taken a southeasterly track more frequently than today due to the blocking effect of persistent and strong anticyclonic patterns over Scandinavia and adjacent areas. Also, he showed that the sub-tropical high pressure belt, which today prevents moist winds from the Atlantic blowing over the Saharan desert region, moved equatorwards. Consequently, with presumably increased rainfall occurring in lower middle latitudes and the subtropics, a humid climate prevailed over regions now affected by drought. It had long been known, as acknowledged by Harmer, that the Saharan region, which today has a virtually rainless climate, formerly experienced comparatively humid conditions during the Pleistocene epoch. Harmer's charts also corroborate the present concept of an expanding and contracting circumpolar vortex of the Northern Hemisphere in cold and warm periods respectively during and since the Pleistocene Ice Age.

It is noteworthy that by graphically indicating prevalent storm tracks and wind directions on his charts, Harmer presented a more dynamic approach to the reconstruction of past circulation patterns than those later made by Lamb. As in geology, these charts demonstrate Harmer's skill in map-making (Figs. 4.1, 4.2, 4.3 and 4.4).

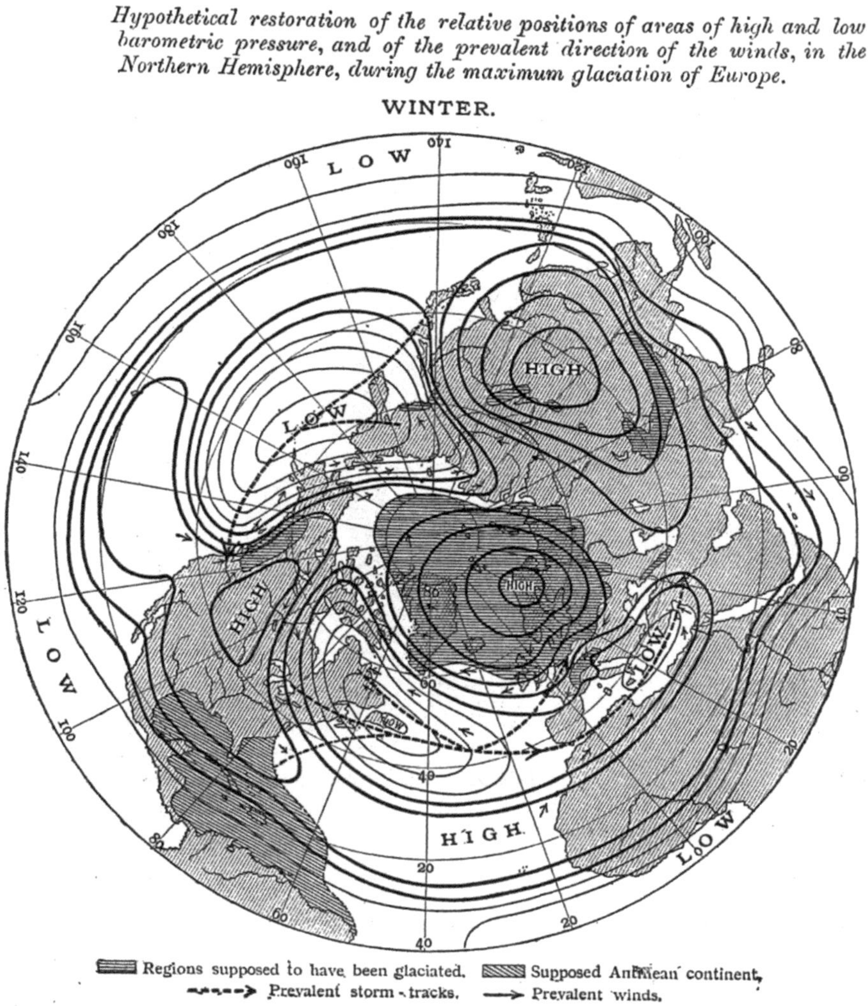

Fig. 4.1 Harmer's reconstruction of the winter surface pressure pattern and prevalent direction of the winds in the Northern Hemisphere during the maximum Pleistocene glaciation of Europe (Harmer 1901)

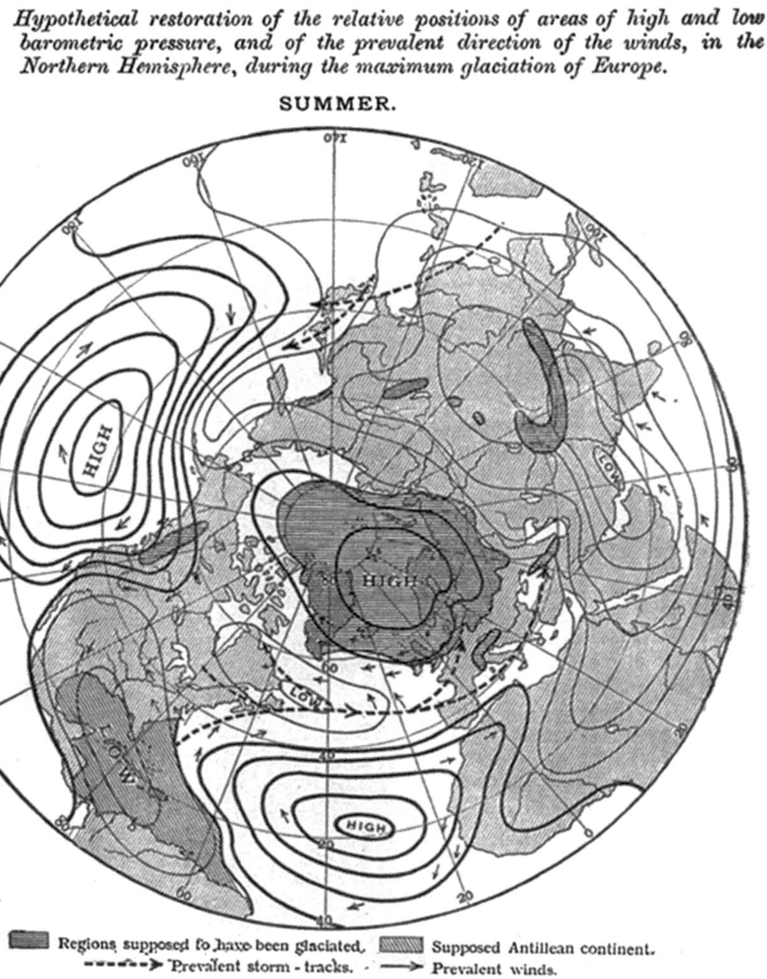

Fig. 4.2 Harmer's reconstruction of the summer surface pressure pattern and prevalent direction of the winds in the Northern Hemisphere during the maximum Pleistocene glaciation of Europe (Harmer 1901)

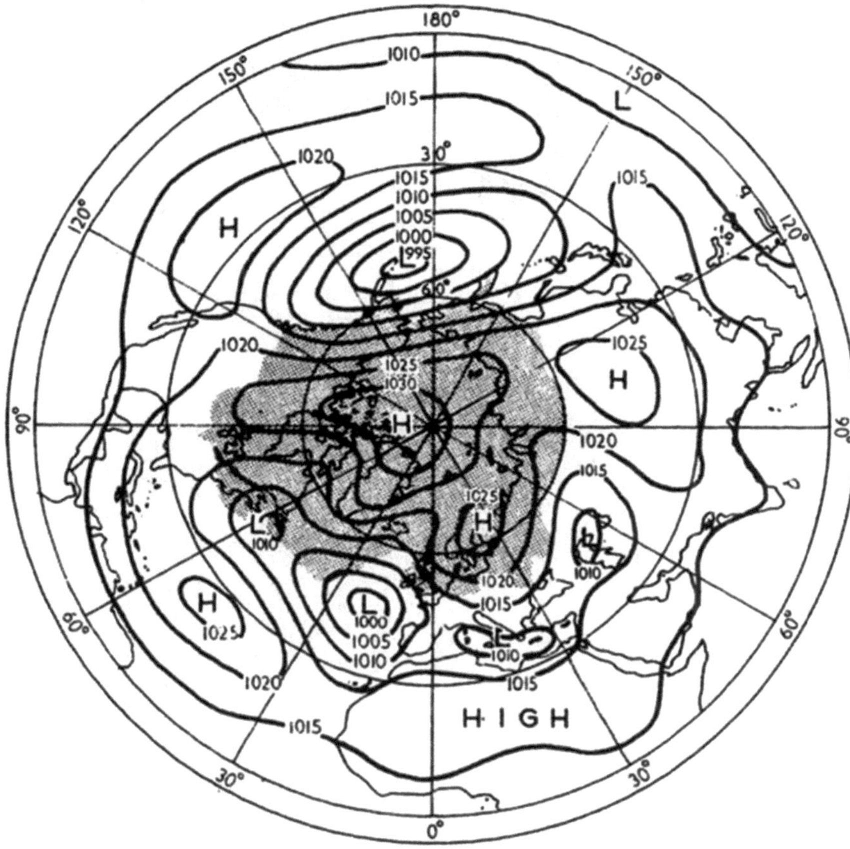

Fig. 4.3 Lamb's reconstruction of the January circulation pattern during the maximum Pleistocene glaciation in the Northern Hemisphere (Lamb 1966)

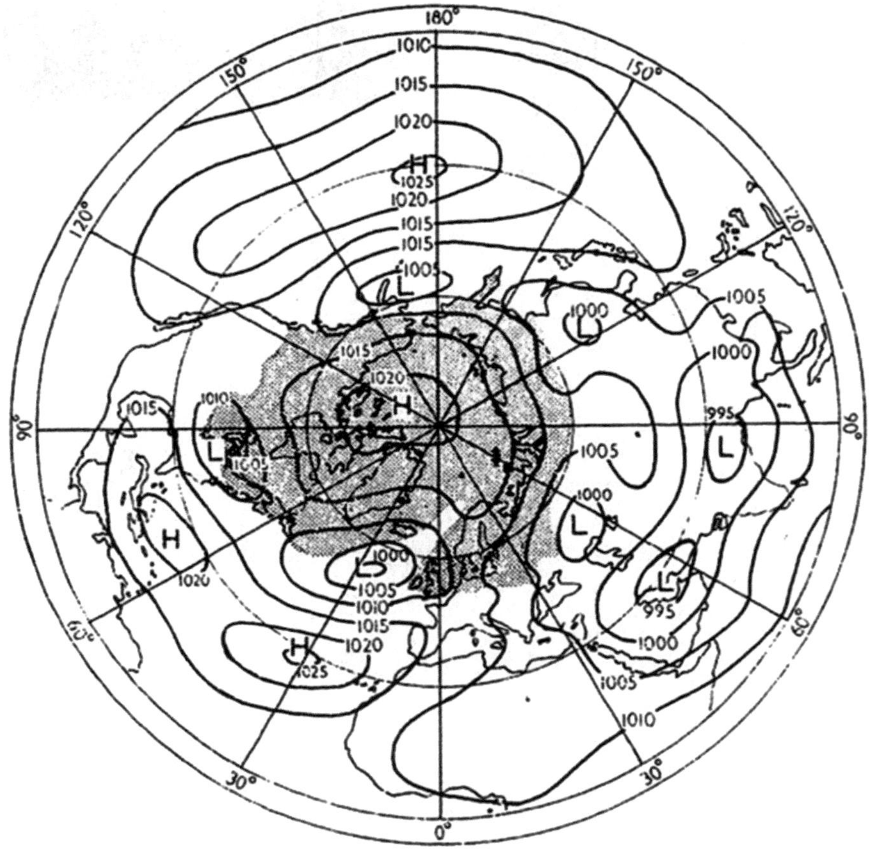

Fig. 4.4 Lamb's reconstruction of the July circulation pattern during the maximum Pleistocene glaciation in the Northern Hemisphere (Lamb 1966)

A comparison of the Harmer and Lamb reconstructions of winter and summer circulation patterns during the maximum Pleistocene glaciation in the Northern Hemisphere brings to light notable points of interest in the two independent attempts to reconstruct past pressure patterns. It was regrettable for the development of climatic research during the 20th century that there had been a hiatus of 60 years between these two studies.

Harmer's analysis of his winter chart shows a large blocking anticyclone over Scandinavia which extended west across Greenland and southwest over the British Isles, resulting in an east to northeasterly wind flow over the latter region; another large anticyclone is situated over Asia. The track of the travelling depressions over the Atlantic divides into two branches mid ocean with one branch proceeding northeast into central Europe and the other east over the Mediterranean. This flow reflects today's finding that a general displacement of the zonal circulation towards lower latitudes occurred during the glacial period and supports the validity of the circulation pattern reconstructed by Harmer.

Lamb's analysis of his winter chart shows an anticyclonic centre over Scandinavia comprising an outlier of the high pressure system over the Arctic basin and like Harmer's chart another large anticyclone is situated over Asia. An extensive belt of low pressure over the Atlantic extends east across the Mediterranean to the Caspian Sea.

Harmer's analysis of his summer chart shows a large blocking anticyclone over the Norwegian Sea covering Scandinavia and much of Greenland, again with an easterly flow over the British Isles. The track of the travelling depressions over the Atlantic now divides into two branches off Iberia with one branch proceeding northeast into central Europe and the other east over the Mediterranean and then via the southern Black Sea into a large area of low pressure over Asia, again probably related to the southward shift of the zonal circulation.

Harmer's charts clearly show that there was less change in the character of the circulation from winter to summer than today, with persistent blocking anticyclonic pressure patterns over Scandinavia effectively controlling the overall situation in both seasons during the maximum glaciation.

Lamb's analysis of his summer chart shows an extensive anticyclone over the Arctic basin with a ridge extending towards northern Scandinavia. A multi-centred area of low pressure is situated over Asia with outlying centres over Europe and off southwest Iberia. Another low pressure system occurs over the eastern North Atlantic.

Lamb concluded that two of the main features of his reconstructed pressure patterns were firstly, an intensification of the circulation, especially in summer, both of the mid latitude westerlies and the Trade winds; and secondly, like Harmer, a general displacement equatorwards of the zonal circulation. Furthermore, he added, it was likely that the winter circulation in higher latitudes over the ice sheets was weaker than today and, following Harmer, that there was less change in the character of the circulation from winter to summer than now.

Harmer stated that the great extension of Swiss glaciers during the Pleistocene epoch indicated that considerable masses of moisture-laden air must have reached the ice fields from maritime sources mostly from the west. Also travelling depressions moving eastward from the Atlantic, as suggested in his summer chart, may have occasionally affected southwestern parts of Europe. He continued that the climate of the region north of the Swiss massif was colder at this period than that to the south of it, the former having been more or less under the influence of anticyclonic winds proceeding from the ice sheets of northern Europe. While ice was steadily accumulating in the north, the sun's heat during summer on the southern slopes of the ice-clad Alpine mountains together with the rains that may have fallen there, produced violent floods which inundated the lowlands of Piedmont and Lombardy from the foot of the Alps to the Adriatic leaving a thick and continuous deposit of diluvial gravel and mud.

> It is only when standing on the tower of the Cathedral at Milan, or on the summit of the Superga, near Turin, looking across the great plain towards the distant mountains rising abruptly from it, that one can realize the strength and volume of the torrents which must have issued from the Alpine valleys at this epoch (Harmer 1901).

One of the effects of strong winds associated with stormy weather over the North Sea is to agitate the water to a considerable depth and to scour the sea-bed, causing molluscan shells to be shifted towards the opposing coast. Harmer learned from a colleague in Great Yarmouth that on 12 or 13 December 1899, the local beach, on which shells were not usually found, was partly covered by them. By referring to the *Daily Weather Reports* published by the Meteorological Office, Harmer found that a spell of strong to gale force east to southeasterly winds had occurred from 7 to 9 December which resulted in the shells being thrown up on the beach.

He used this meteorological situation, associated with a blocking anticyclone over Scandinavia as an analogue for the type of circulation pattern which, he suggested, had frequently occurred during the late Pliocene leading to the formation of a succession of littoral beds, which, in the case of the Red Crag, attained a thickness of up to 6 m (20 ft) in the cliffs of East Anglia.

In summary, Harmer's hypothesis suggested that anomalous circulation patterns associated with current weather extremes over the British Isles may have been the norm in past geological periods such as the late Pliocene epoch when shelly marine sands, collectively known as Crag deposits had massively accumulated near the western shoreline of the Crag Sea, due to the successive drifting of the material upon the beach by strong and persistent easterly winds.

In 1925, 2 years after Harmer's death, his second key paper on the meteorological conditions that prevailed during the Pleistocene epoch, 'Further Remarks on the Meteorological Conditions of the Pleistocene Epoch', was presented by the pioneering climatologist with expertise in geology, Dr Charles E.P. Brooks (1888–1957) at a meeting of the Royal Meteorological Society held on Wednesday, 20 May, at the Society's Rooms, 49 Cromwell Road, South Kensington, Mr J.S. Dines, Vice-President, being in the Chair.

Brooks explained that this paper was essentially an amplification and extension of the views first put forward by Harmer in his 1901 paper, 'The influence of the winds upon climate during the Pleistocene Epoch: A palaeometeorological explanation of some geological problems'. Harmer's son, Sir Sidney Harmer, who attended the meeting, explained that he had found the draft manuscript in his father's collection of papers and was greatly indebted to Brooks for preparing the article for publication (Brooks 1970).

In his introductory note, Brooks stated that the manuscript had been left by Harmer in an almost complete form. Although some of the maps had not been drawn and the text had not been finally revised, the material was otherwise complete. However, on reading it through, Brooks found that a certain problem arose which he felt should be addressed before publication.

Brooks stated that while most of Harmer's deductions were still valid in the light of current knowledge, in one respect they have proved to be erroneous. The hypothesis of alternating glaciations in Europe and North America, which was contained in Harmer's 1901 paper, had been shown to be untenable due to research by the Swedish geologist, Gerard De Geer (1858–1943) and other workers on varves (distinct layers of sediments deposited annually in glacial lakes); the latest glacial stage was certainly contemporaneous on both sides of the North Atlantic. As Harmer would no doubt have accepted this view, Brooks thought it best to delete any passages which depended on this supposed alternation of glaciations. In other respects the text was practically unchanged.

In supplying a figure that was missing from the manuscript, Brooks endeavoured to interpret Harmer's views as depicted in those of his 1901 paper, showing surface pressure patterns and prevailing wind directions in the Northern Hemisphere during the maximum Pleistocene glaciation of Europe. Brooks mentioned that the construction of this new figure required some explanation. Its purpose was to show how by elevation a ridge of land was formed connecting Greenland with Europe. Although Harmer had supposed that at the same time the northern part of North America was depressed below its present level, Brooks noted that this was probably not the case and accordingly suggested the following compromise: between Greenland and Scandinavia the land was supposed to have extended to the submarine contour of 1,000 m (3,300 ft) below sea-level; on the west coast of Greenland, the east and north coasts of North America, the American Arctic Archipelago, the British Isles and France, the submarine contour of 200 m (660 ft) below sea-level has been taken as the limit of the former land extension. Elsewhere (including Svalbard) no change was made, except that Brooks followed Harmer in indicating the existence of an Antillean extension of the North American continental area.

Harmer began his paper by stating that when the distinguished geologist, Sir Charles Lyell (1797–1875) discussed the possible effect of astronomical causes on past climates (particularly that of the Pleistocene epoch) in the 10th edition of his *Principles of Geology* (1868), he reverted to the theory he had proposed over 30 years earlier in the first edition of his volume, that it is only by admitting the frequent and paramount importance of geographical changes on the earth itself (such as variations in the height and position of the land and the course of ocean currents) that those great revolutions of climate can be explained 'which wrapt large portions of the Northern Hemisphere in a winding sheet of continental ice' (Fleming 1998).

Harmer would also have been aware that in his *Principles of Geology* Lyell stated that in the present period the oceans have free communication over the whole globe; as a result the warm waters of the equator can penetrate to the poles and the annually accumulated ice of the Arctic regions is melted by such warm southern currents. But suppose, Lyell suggested, the Arctic and north temperate sea were completely, or almost enclosed by land, as might easily occur by a comparatively limited elevation of the sea bed between Greenland, Iceland and the Orkney Islands, and at the entrance to the Davis Strait, would not the ice, he queried, of each successive winter accumulate so as to form a vast ice cap which would inevitably produce a glacial epoch over the whole north temperate zone. Therefore changes in the distribution of sea and land are a sufficient cause, Lyell believed, for any variations of climate of which we have geological evidence. That changes of the required amount must certainly have occurred at various epochs of past time is also certain. On the other hand whilst Lyell presented a very strong argument against these changes having been the sole cause of those variations of climate of which there is such clear evidence, he further stated that they may have always been an important and perhaps an essential collateral agent in their production. His two 'ideal' maps illustrating conditions of the earth which, by geographical changes alone, would bring about a perpetual summer or an almost universal winter provide models for later reconstructions of past climates (Wallace 1895).

Harmer continued that among the various geographical hypotheses which had been suggested concerning the probable effect of the closing of the gap between Greenland and Europe upon the North Atlantic Drift that of Alfred R. Wallace (1823–1913) should also be especially mentioned. Although Harmer had referred to this idea in his 1901 paper he believed it would be valuable to re-consider this concept in greater detail, especially from a meteorological standpoint.

Wallace was the founder of biogeography and his enduring legacy in this subject is the 'Wallace Line', a faunal barrier separating the species of Asia from those of Australia. This invisible boundary catalyzed Wallace's theory of evolution and prompted Charles Darwin to articulate his own theory.

Wallace had stated that despite certain criticisms made about the views of Lyell whose fundamental theory of climatic change was based on changes in the distribution of land and sea, especially in polar and equatorial regions, he (Wallace) believed that Lyell was substantially correct and that the global maps included in his *Principles of Geology* showed 'a condition of the earth which, by geographical changes alone, would bring about a perpetual summer or an almost universal winter' (Wallace 1880).

As Harmer explained in his 1925 paper, the North Atlantic region is an area of paramount interest and importance to glaciologists since the more striking phenomena of the Ice Age had been grouped around it:

> The physiographical conditions of the North Atlantic basin are unique. There only in the Northern Hemisphere does uninterrupted communication exist between the equatorial and polar oceans. Through the more easterly portions of the great channel, 600 miles wide, which separates Greenland from the Eurasian continent, warm surface currents from the tropics carry their ameliorating influence to the north of the Arctic Circle, while to the west of Iceland cold water flows southward through the Denmark Strait, and from the Davis Strait by means of the Labrador Current towards North America. But this is not all; the existence of this channel offers an unobstructed passage northeastwards for the Atlantic cyclones, many of which find their way, especially in winter, by this route to the North Cape and beyond, the result being that the British Isles and the Norwegian coast are at that season under the prevalent influence of warm southwesterly winds, while Labrador and New England are flooded by aerial currents from the ice sheet of Greenland, and the still colder regions of the Arctic Archipelago, over which the average winter temperature ranges from −20 to −30 °F. The aerial are possibly of more direct importance than the ocean currents, as they are certainly more rapid in their action. Water from the Labrador Current is carried eastward by the westerly winds. Northward movement of water (the Irminger Current) takes place west of Iceland and west of Greenland. The Norwegian current of the Gulf Stream is by far the largest and melts an enormous amount of ice (Harmer 1925).

The climate, he continued, is abnormal on both sides of the North Atlantic. To the west, for example in Labrador, the −18 °C (0 °F) January isotherm passes a little to the north of the 50th parallel. To the east, the same isotherm reaches Spitzbergen (Svalbard) in March and April just to the south of latitude 80° N. In the North Pacific, on the other hand, where little interchange of heated and chilled water can take place through the shallow and restricted opening of the Bering Strait and where the direction of the prevalent winds is of a different character, there is less variation between the average temperature of the coasts of North America and of Asia in similar latitudes, either in winter or summer. Harmer continued the story:

> It may be instructive, therefore, to compare the present conditions, geographical and meteorological, of the basins of the North Atlantic and the North Pacific. Geographically, these regions closely resemble each other, except in one respect. Both are similarly open to the south, but the latter is (practically) closed to the north. Communication exists, however, between the North Atlantic and the Polar Sea not only through the Icelandic Channel, but also by means of the Davis Strait and fiords of the Arctic Archipelago. At present the latter are blocked by the ice pack, but at a certain stage of the glacial epoch they may have exerted an important influence on the climate of North America.

It seems clear that the closing of the gap between Greenland and Europe would effect a change in the winter temperatures of northwestern Europe by terminating the geographical conditions to which its normally mild climate is due. Such an exclusion of the warm waters of the North Atlantic, as described by Wallace, would result in an extraordinary accumulation of snow and ice on the mountains of Scandinavia and Britain:

> In the region with which we are most immediately interested it is easy to see how a comparatively slight alteration of land and sea, such as has undoubtedly occurred, would produce an enormous effect on climate. Let us suppose, for instance, that the British Isles again became continental, and that this continental land extended across the Färoe Islands and Iceland to Greenland. The whole of the warm waters of the Atlantic, with the Gulf Stream, would then be shut out from Northern Europe, and the result would almost certainly be that snow would accumulate on the high mountains of Scandinavia till they became glaciated to as great extent as Greenland, and the cold thus produced would react on our own country and cover the Grampians with perpetual snow (Wallace 1880).

However, Harmer believed that the diversion of the storm tracks and the consequent alteration in the direction of the prevailing winds would have been an even more important factor in such a situation. The northward track of Atlantic depressions, as well as that of the North Atlantic Drift, would be barred by the existence of an Icelandic bridge, the former would be turned towards the southeast, as those are now in the north Pacific.

As he mentioned, the displacement of the mean winter low pressure area of the North Atlantic to the south of the supposed barrier would certainly produce changes of far-reaching importance in the atmospheric circulation as well as that of the ocean waters (Fig. 4.5):

An interesting argument in favour of the existence of the Icelandic bridge during the maximum glaciation of Scandinavia and Great Britain is furnished by the difference in the direction of the ice movement in those countries. In the latter the thickness of the ice, and consequently the amount of precipitation, is said to have been greatest in the west; the ice, for example, which carried the shap boulders from Westmorland to the Yorkshire coast, similarly crossed the Pennine watershed from west to east. In Scandinavia, on the other hand, the movement was in the opposite direction, from Sweden towards the west; the precipitation must therefore have been greater in the Baltic region than on the Norwegian coast. The explanation of these facts may be that the western part of the British Isles was at that time in close contiguity with the Atlantic, and under the influence of some of its cyclonic disturbances. The moisture bearing cyclones, on the other hand, were prevented from reaching the Norwegian coast by the Icelandic barrier and the ice-fields to the north consequently resulting, but a path towards the Baltic lay open to them across northern France where, the rainfall was then excessive. How this may have taken place is shown hypothetically (Harmer 1925).

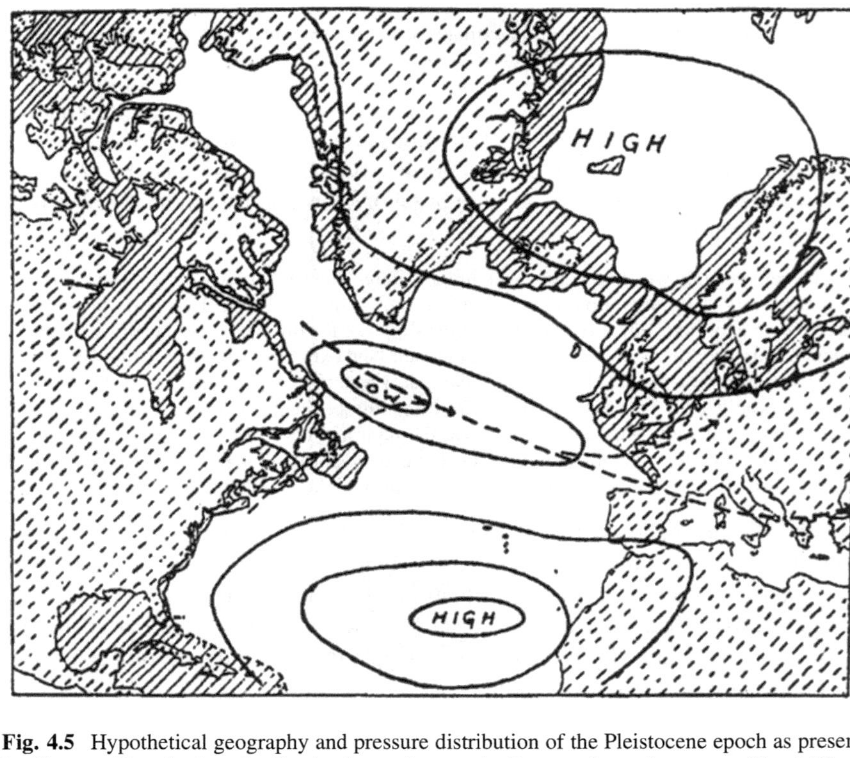

Fig. 4.5 Hypothetical geography and pressure distribution of the Pleistocene epoch as presented by Harmer. Key: *broken shading* land; *continuous shading* sea formerly proposed land (Harmer 1925)

In his 1901 paper, Harmer had suggested that some well-known case studies of anomalous past climates, such as the existence during the Pleistocene epoch of humid conditions in the Sahara (a region now arid) might be explained meteorologically by supposing that the distribution of high and low pressure areas, and the prevalent tracks of Atlantic depressions were different to those which now prevail, and that such a situation was not only contemporaneous with but also due to the existence of the great ice sheets of North America and Europe.

Harmer referred to an article by the meteorologist, Henry N. Dickson (1866–1922) concerning the mean temperature of the atmosphere and causes of glacial periods also published in 1901. While accepting the view that the case study of the Sahara and other similar geological anomalies might be explained meteorologically, Dickson argued that the shifting of the Atlantic storm tracks towards the south during the glacial period was due to the increased difference in temperature then existing between the polar and equatorial regions. Refrigeration having been, in his opinion, considerably more in the former than in the latter, the steepness of the thermal gradient would be greater than at present and this would have increased the wind force of the North-East Trades and the mid-latitude Westerlies, and displaced the sub-tropical high pressure belts to latitudes nearer the equator. This increased vigour of the circulation would tend the development of cyclonic pressure systems, the stronger the flow the more intense and frequent these depressions would have become; also the 'normal' storm tracks would have been displaced to lower latitudes, running more from west to east rather than from southwest to northeast—as at present.

Dickson had studied at Edinburgh University where, among other projects, he acted as a volunteer assistant at the Ben Nevis Observatory. Harmer well acknowledged Dickson 'from whom I have received some friendly and valuable criticism'. Dickson's paper, 'Mean temperature of the atmosphere and the causes of glacial periods' was read at a Meeting of the British Association and published in *The Geographical Journal* for November 1901.

> Starting with the assumption that secular variations of climate in the past have been due to changes in the mean temperature of the atmosphere, the author [Dickson] pointed out that such changes had probably been associated with large alterations in the temperature gradient between the equator and the poles. But as this difference of temperature is the primary cause of the overall planetary circulation of the atmosphere, the form and intensity of the latter must have varied with it. Thus a lowering of the mean temperature [with relatively greater cooling at the poles] would be accompanied by an increase in the equator-poleward gradient, a rise by a diminution of it, and in the former case the planetary circulation would become more active and the sub-tropical high pressure belts would be displaced to lower latitudes. It is suggested that the effect of these changes on the distribution of precipitation and on the position and directions of the major storm tracks, may explain some peculiar features of glacial phenomena (Dickson 1902).

Harmer continued that conditions not only of greatly increased rainfall but also of torrential downpours must have existed in southern Europe during the glacial epoch (as had already suggested in his 1901 article) and these effects may have been due to the greater strength of the general circulation together with the increased temperature gradient. At present the latter is, in the Northern Hemisphere, much steeper during the winter than during the summer months. For example, in January the difference between the average temperature of the North Pole and the Equator is 115 °F (−35 °F in the first case and 80 °F in the second); in July it is only 55 °F (35 °F and 90 °F, respectively).

It seems that, like today, storm tracks in the Pleistocene epoch may have been determined by the prevalent position of the anticyclones of that period and especially by those high pressure systems which may have prevailed over the region occupied by great ice sheets. Moreover, it seems possible, that the difference in the temperature gradients of the Pleistocene epoch and the present day may not have been as great as that now existing between those of January and July, and that the increased steepness may have been local rather than general. Over the European ice sheet, for example, the gradient may not have been greater than that which now prevails in winter over Greenland and the Arctic Ocean. Harmer continued:

> Dealing, however, with the maximum glaciation of Europe it would appear that it was over the region between the southern edge of the European ice and the Mediterranean that the crowding together of the isotherms might have been greatest, and it is there that the phenomenal erosion of the Pliocene and post-Pliocene strata has taken place. The sharp manner in which the glaciers descending the southern valleys of the Alps were cut off when reaching the great plain of Piedmont shows, I think, that the climate of Northern Italy at this epoch was not severe. There would then have been a strongly marked thermal gradient between Spitzbergen [Svalbard] and the Mediterranean, similar to that which now exists in North America, but as before pointed out, its steepest part would lie to the south of the European ice-sheet.
>
> Under such circumstances, storm-tracks would have probably crossed the Atlantic in winter, moving eastwards as they do now, but their course northwards being blocked, their greatest fury would be expended over Central and Southern Europe, resulting in an increased accumulation of snow upon the mountain regions of the Pyrenees and the Alps, and torrential rains over the lower ground. The influence of the ice-sheet, close at hand, would cause snow to fall upon the mountain regions during the summer months at elevations where precipitation now takes place in the form of rain.

In a review of the characteristics of the oceanic and atmospheric circulations in the Northern Hemisphere, Harmer stated that the chief difference between the North Atlantic and North Pacific Oceans is that the former is open to the north, while the latter is practically closed to the north, since the Behring Strait is too narrow and shallow to allow the passage of an appreciable current. Hence in the North Atlantic the North Atlantic Drift as a broad extension of the Gulf Stream flows northeastward into the Arctic Ocean, while in the Pacific Ocean the Kuroshio or Japanese Current is forced to turn south-eastward along the coast of North America. This dissimilarity causes differences in the pressure distribution together with differences in both oceanic and atmospheric circulations which combine to give Western Europe a more genial climate than the west of North America.

From this comparison Harmer concluded that the closing of the channel between Greenland and Europe would bring about changes in the oceanic and atmospheric circulations which would suffice to cause a glacial epoch in Europe. The diversion of storm tracks and consequent alteration in the direction of the prevailing winds are probably even more important than the changes in the ocean currents. Finally, Harmer dealt with the climatic changes that occurred in the Mediterranean region during the glacial period and argued that the increased temperature gradient in southern Europe caused a notable increase of storminess in that region which resulted in torrential rains for which verifiable evidence is available.

In the discussion which followed the reading of Harmer's 1925 paper, his son, Sir Sidney Harmer, said he was greatly obliged to the Royal Meteorological Society for asking him to speak, but he must entirely disclaim being a meteorologist. Although he had no scientific contribution to make to the discussion, he greatly appreciated the acceptance by this Society of his father's last paper. It was about a subject in which, as he knew from personal experience, his father took a very great interest from the time when his first paper was published in 1901. The manuscript was found after Harmer's death amongst a large number of other manuscripts, most of them incomplete drafts of work in prospect and it had been found impossible to make use of the majority of them. This paper, however, was in a more complete and consecutive form and it seemed possible that it might be considered worthy of publication. Sir Sidney was greatly indebted to Mr Brooks, who had made very few alterations to the paper, preparing two figures which had obviously been contemplated and cutting out certain parts which had to be omitted in view of more recent information on the subject, but had practically left the paper as Harmer wrote it. There was, he thought, very little doubt that it was intended to complete the paper, but a large and arduous piece of work, the monograph on the Pliocene mollusca of Britain, took up the whole of the remainder of his father's working years and was practically finished at his death. He, Sir Sidney, felt that his father, if he had lived, would have come back to the subject of the present paper with the intention of communicating it to the Society.

Sir William Napier Shaw, F.R.S. (1854–1945), Director of the Meteorological Office from 1905 to 1920, said he was glad to be present because he had often discussed the subject with Harmer when the latter visited the Meteorological Office, situated then in Victoria Street, London. Previously, Harmer had expressed his great indebtedness to Shaw who not only had allowed him to make constant use of the valuable library at the Meteorological Office but had been kind enough to give him the benefit of his (Shaw's) experience and of making some important suggestions.

Shaw continued that Harmer used to lay stress on the fact that formerly shells were deposited in great abundance on the east coast of England, whereas now they were not at all abundant in this region. Harmer suggested that it was a change in the condition of the earth and that the cessation of the shell deposits was due to a variation in the meteorological conditions. In former times, he stated, the beat of the waves and the mean strength and direction of the wind were towards the east coast and brought the refuse of the sea there instead of, as now, leaving sand behind and clearing away the refuse into the North Sea. The wind must have been from the east when the fossils were deposited whereas now the prevailing wind is from the west. Then comes this question: if there were a bar across the North Atlantic which brought land further south by 500 miles or more, would that so alter the meteorological conditions as to give easterly winds on the east coast of

England and what would be the effect from a geological point of view? If the meteorological conditions were altered in such a way, should we get a prevailing easterly wind? The question interested Harmer much at the time when it was put forward and he thought it a very useful form of speculation. It started a subject which had proved to be one of the most attractive sections of meteorological inquiry. In America they were devoting much attention to palaeometeorological work: that is to say, the meteorology associated with changed geological conditions. Brooks himself, who has done not a little work of that sort in this country, was associated with Harmer in the inquiry at the Meteorological Office and from that has developed a large amount of literature which is both geological and meteorological. It is impossible to separate the geological from the meteorological, as the two are expressions of the results of the same forces. It gave him much pleasure to hear Harmer's paper read before the Society and to know that it will find a permanent home in the Society's records.

In 1940 the palaeometeorological research of Harmer was again presented to the Royal Meteorological Society this time by the leading British geologist Professor Percy G.H. Boswell, F.R.S. in the George J. Symons Memorial Lecture on 17 April that year.

> In connection with the problem of climatic fluctuations during the Ice Age, it is desirable to refer to one of the few papers dealing with palaeometeorology that have appeared in geological journals. In 1901, my old friend and master, the late F.W. Harmer, contributed a paper to the Quarterly Journal of the Geological Society on the influence of the winds upon climate during past epochs. I mention it as the kind of speculation upon which geologists may embark, although very few have the temerity to do so (Boswell 1940).

Boswell called attention to the general principle that anomalous meteorological situations in the past, even if more or less permanent, may have been due to the same cause which brings about temporary changes of a similar character in the present period, namely, variations in circulation patterns. Boswell continued that Harmer appeared to have been drawn to consider this concept through his studies of the Pliocene shell beds, or Crags of East Anglia. In his efforts to interpret the conditions under which these shallow water deposits were formed, Boswell reiterated that Harmer referred to similar deposits being accumulated in the present period on the Dutch coasts, especially when driven up on the beaches by strong southwesterly winds. As a result, Harmer concluded that strong easterly winds had occurred instead during the Pliocene epoch, causing shells and other marine organisms to be torn from the sea bed and thrown up as detrimental accumulations on the eastern England shores of the proto-North Sea.

Boswell continued that Harmer concerned himself particularly with the conditions just before and during the Ice Age, and called attention to the general principle that anomalous weather in the past, even if more or less permanent, may have been due to the same cause which brings about temporary changes of a similar character at present, namely, a change in the direction of the wind. He seems to have been led to the consideration of the problem by his studies of the Pliocene shell beds, or Crags, of his East Anglian homeland. In his efforts to interpret the conditions under which these world famous shallow water deposits were formed, he called to mind the similar deposits of shelly sands now being accumulated on the Dutch coasts especially when driven up by south-westerly gales. Thus he was brought to a belief in the existence of easterly gales during the Pliocene epoch, under the action of which shells and other marine organisms were torn from the sea-bed and thrown up as detrital accumulations on the East Anglian shores of the North Sea of those times' (Boswell 1940).

4.4 Conclusion

In view of the herewith presented research by Harmer it was an unfortunate loss to the development of atmospheric science that his novel approach to palaeometeorology linking past changes of climate with variations in the general circulation, as depicted by mapping methods, was not followed up during the early years of the 20th century and that it was only in the 1960s that the renowned climatologist, Professor Hubert H. Lamb (1913–1997), first in the Meteorological Office and later in the Climatic Research Unit, reintroduced the concept in order to obtain a better understanding and knowledge of climatic change (Lamb 1966).

References

Boswell, P.G.H., 1940: "Climates of the Past: A review of the Geological Evidence", in: *Quarterly Journal of the Royal Meteorological Society*, 66: 249–274.

Brooks, C.E.P., 1970: *Climate Through the Ages: A Study of the Climatic Factors and Their Variations* (New York: Dover Publications).

Dickson, H.N., 1902: "Mean Temperature of the Atmosphere and the Causes of Glacial Periods", in: *Quarterly Journal of the Royal Meteorological Society*, 28: 32.

Fleming, J.R., 1998: Charles Lyell and Climate: Speculation and Certainty (London: Special Publications, Geological Society): 143: 161–169

Harmer, F.W., 1901: "The Influence of the Winds upon Climate During the Pleistocene Epoch: A palaeometeorological Explanation of Some Geological Problems", in: *Quarterly Journal of the Geological Society*, 57: 405–476.

Harmer, F.W., 1902: "A Sketch of the Later Tertiary History of East Anglia", in: *Proceedings of the Geologists' Association*, 17: 416–451.

Harmer, F.W., 1925 (posthumously): "Further Remarks on the Meteorological Conditions of the Pleistocene Epoch", in: *Quarterly Journal of the Royal Meteorological Society*, 51: 247–259.

Kington, J.A., 2010: *Climate and Weather* (London: New Naturalist Library, Collins).

Lamb, H.H., 1966: *The Changing Climate* (London: Methuen).

Wallace, A.R., 1895: *Island Life or the Phenomena and Causes of Insular Faunas and Floras*, Second and revised Edn. (London: Macmillan).

Chapter 5
Final Years

Abstract Visit by Professor Kendall to Harmer in his late 80s and their concern about the loss of geological field sites due to increasing man-made activities. Reference to Harmer's last two outstanding published works: a map of Pleistocene boulder-clay types and erratic trails, and a monograph on Pliocene mollusca. History of the Crag Sea, Doggerland, mollusca as indicators of past climatic conditions and the re-classification of the Red Crag in the geological table.

Keywords Geological field sites · Boulder-clay · Erratics · Pliocene · Pleistocene · Mollusca · East Anglian Crags · Crag Sea · Doggerland

Following his father's death in 1923, Sir Sidney Harmer when writing to a colleague stated:

> In his later years my father had been cut off from his scientific friends, but it was always a satisfaction to him to know that his work was appreciated. His work was practically completed, because although there is still something to be published by the Palaeontographical Society he had passed for press the Part which is to appear in due course, and he had completed the MSS, with one Plate, of what he had intended to be actually the final instalment [Volume II, 1925]. His scientific work was a great resource to him, and he was engaged with it almost to the end.

Thus Harmer in his late eighties would have been pleased that in the summer of 1922 one of his esteemed colleagues, the geologist Professor Percy Kendall (1856–1936) made 'a long desired visit to my old friend, F.W. Harmer. I found him in delicate health, as was not surprising in view of the fact that he was then in his 88th year'. Harmer was particularly happy to meet his friend again as earlier Kendall had handed over the results of his own research on Waltonian Red Crag fauna to Harmer with regard to his friend's work on the same formation at the Little Oakley site.

Indeed, one of the subjects that arose during their conversation was the increasing loss of geological field sites; Kendall remarked that increased industrial activities for road-making and building materials, especially the concentration of

J. A. Kington, *Frederic W. Harmer: A Scientific Biography*, SpringerBriefs in Environment, Security, Development and Peace 19, DOI: 10.1007/978-3-319-07704-8_5,

large brick-works on outcrops of argillaceous formations, had resulted in the closure of hundreds of small excavations formerly used for geological digs, such as gravel-pits, chalk-pits and brickyards. Harmer too commented regretfully on the disappearance of such exposures for geological field studies.

Two outstanding works undertaken by Harmer during the final years of his life, each entailing an immense amount of research, were, firstly, a detailed map showing the types of boulder-clay and trails of erratics in England and Wales, and secondly a two-volume monograph on Pliocene Mollusca published by the Palaeontographical Society. This latter work, his *magnum opus*, was an achievement which will long earn the gratitude of later investigators and will always remain a most fitting monument to his memory.

Harmer stated that at the Little Oakley site about 200 tons of material had been sifted and examined. He added that to prevent disappointment to any persons who might wish to visit this prolific locality, he should mention that by arrangement with the owner of the Oakley estate, the excavations were filled up and levelled down as the work proceeded. Harmer had proposed to offer his collection of fossils from this site to the Sedgwick Museum in Cambridge (Harmer 1914–1919).

5.1 Volume I

In 1913, writing from his home in Cringleford, near Norwich Harmer included the following notable comments in his Introduction:

> More than sixty years have elapsed since the publication of the classical Monograph on the Mollusca of the Crag by my old friend, Searles V. Wood, and between thirty to forty years since that of the Supplements to it. Much fresh material it seems desirable to describe has been obtained, especially from a new and interesting section at the village of Little Oakley, near Harwich, midway between Walton-on-Naze and Felixstowe.
>
> Many important works dealing with the Pliocene and Pleistocene, as well as with the Recent Mollusca of various parts of the northern hemisphere, have appeared, moreover, during recent years.
>
> The continued study of the Crag beds is not only desirable, but likely to prove of great interest and importance. Although isolated and fragmentary records of the Pliocene history occur at Lenham [Kent], St. Erth [Cornwall], in the Cotentin [Normandy], in north-east Scotland, and probably in the Isle of Man, it is only in the Anglo-Belgian basin and the little-known Crag of Iceland that we have a more or less connected series of fossiliferous deposits from which we may ascertain the character of the molluscan fauna of the seas of north-western Europe [and by inference past climatic conditions] during the period intervening between the Miocene and Pleistocene epochs. No older Pliocene strata are known in Scandinavia, any which may once have existed in that region having been destroyed by the erosion of the great ice sheets.

5.2 Volume II

In September 1924, due to Harmer's death the previous year, Alfred Bell of the Ipswich Museum undertook the writing of the Preface to this Volume in which he included the following remarks:

> The late author of this Monograph realized his wish to complete his account of the Pliocene Gasteropoda before his death, which occurred on April 11th, 1923. Part III of his second volume had been passed by him for press several months before its publication, and the fourth and concluding part was also ready. An Index to the second volume had been compiled by him, but Sir Arthur Smith Woodward has kindly undertaken to have this work done independently, in order to assure the accuracy which is so important in an Index.
>
> Mr. Harmer had intended to express his sincere thanks to all those friends at home and abroad who had helped him in difficulties by the loan or gift of specimens for comparison, or by the determination of critical forms. This duty he left for his son, Sir Sidney Harmer, and myself to carry out, in his name and for him.

5.3 General Remarks

In geology, Crag is the collective term given to the sandy, shelly Pliocene deposits which comprise the principal Tertiaries of East Anglia laid down about 2 million years ago under marine and later estuarine conditions. Patches of the Red Crag have survived in northeast Essex at Dovercourt, Beaumont, Wrabness and Little Oakley. In the 19th and early 20th centuries sand pits at these sites yielded fossils, one of the most well known sites was a shallow pit at Little Oakley that was investigated by Harmer who, in his later years, sifted and examined its sand for fossil shells. As a result, 650 different species of mollusca and about 100 specimens of polyzoa were found in this one pit alone (nearly 400 of the species illustrated in his monograph were from the Little Oakley pit).

Harmer's meticulous research has shown the extraordinarily rich mollusca fauna of the Crag Sea which existed over northeast Essex and East Anglia about 2 million years ago. He was aware that the Red Crag formation was deposited in near-shore conditions during the Pliocene as part of a large delta that was building out into the southern part of the Crag Sea which, at that time, extended into the London Basin as far west as Hertfordshire, where a proto-Thames flowed into it. Although this Crag Sea or proto-North Sea was connected to the Atlantic across southern Britain it was not through the English Channel as it is known today. However, this early sea link was slowly reduced and by about 2 million years ago not only was it lost but the head of the delta between Britain and mainland Europe became dry land, now named Doggerland (after the Dogger Bank). As the shoreline moved northwards, so the proto-Thames and other major rivers such as the Rhine flowing into the Crag Sea extended across the former sea bed. The molluscan shells of the Red Crag indicate a cooling of the climate compared to previous stages and this formation is sometimes considered to mark the onset of the last Ice Age.

This is in accordance with the re-classification since Harmer's time of the East Anglian Crags in the geological column, whereby it is now generally agreed that whilst the Coralline Crag should remain in the Pliocene Epoch, the Red Crag and Norwich Crag should be designated to the Pleistocene (Challinor 1978). Thus today the division between the Pliocene and the Pleistocene epochs in Britain is placed at the boundary between the Coralline Crag and the Red Crag of East Anglia. At this point there is a relatively clear stratigraphical break, a marked increase in the proportion of modern forms of marine mollusca and mollusca of northern aspect, as well as the first arrival (in the Red Crag) of the antecedents of today's fauna species such as the elephant and horse.

References

Challinor, J., 1978: *A Dictionary of Geology* (Cardiff: University of Wales Press).

Harmer, F.W., 1914–1919: *A Monograph of the Pliocene Mollusca of Great Britain*, being supplementary to Searles V. Wood's *Monograph of the Crag Mollusca*, Volume I, (London: Palaeontographical Society).

Glossary

Amstelian former term named after the small Dutch river Amstel marine beds of this Pliocene stage were first discovered in a borehole near Amsterdam

Argillaceous term applied to rocks which comprise grain-sized sediments less than 0.0625 mm (0.0025 in) in diameter; they include over 50 % of sedimentary rocks with most having substantial clay mineral content

Boulder-clay fine-grained glacial deposit formed by the grinding action of ice over a land surface

Butleyan Crag the Red Crag of the Butley region of Suffolk

Cannon-shot gravel deposit comprising bands of large nodules of hard, dark grey, argillaceous gritstone

Casterlien Pliocene sandstone deposit found in Belgium

Chillesford Crag deposit of the Icenian Crag beds

Contorted drift glacial deposit in East Anglian cliffs especially near Cromer; former ice sheet movements resulted in the enclosed deposits being strongly folded and when the ice eventually melted this material remained deformed in situ

Coralline Crag Pliocene deposit lowest of the East Anglian Crags

Crag Sea great gulf between Britain and mainland Europe during the Pliocene

Crags of East Anglia shelly sandstone strata divided stratigraphical (from early to late) into the Coralline Crag, Red Crag and Norwich Crag

Cretaceous third and most recent of the three Periods comprising the Mesozoic Era (146–65.5 million years ago)

Cromer Forest Beds Series estuarine and fresh-water deposits found in the Norfolk coast inferred past climatic conditions varied from cold to temperate

Cromer Till boulder-clay deposit containing Scandinavian erratics

J. A. Kington, *Frederic W. Harmer: A Scientific Biography*, SpringerBriefs in Environment, Security, Development and Peace 19, DOI: 10.1007/978-3-319-07704-8,

Diestien Beds Pliocene sandstone deposits found in Belgium comparable to Lenham Beds on the English side of the Dover Straight

Diluvium term applied during the early 19th century to extensive superficial deposits supposed to have been due to a biblical deluge such as Noah's Flood. These deposits are now known to be mostly glacial drift

Dip inclination of a rock stratum at right angles to the strike

Drift sediment deposited or related to glacial ice activity

East Anglian Crags comprising three deposits: the Coralline Crag, Red Crag and Norwich Crag. Although formerly all three were put in the Pliocene Epoch only the Coralline Crag is so placed today (the other two Crags are classed as Pleistocene)

Erratic glacially transported rock whose lithology shows that it could not have been eroded from local rock. Glacial erratics found in East Anglia include rock types from Scandinavia

Gasteropoda class of mollusc with locomotive organ placed ventrally

Icenian Crag comprises the Norwich Crag, Chillesford Beds and Weybourne Crag, shallow-water marine and estuarine deposits

Jurassic one of the three Periods comprising the Mesozoic Era (199.6–146 million years ago)

Kettle hole depression in the surface of glacial drift, resulting from the melting of an included ice mass

Kimmeridge Clay deposit of the Kimmeridgian stage of the upper Jurassic (156–151 million years ago)

Lenham Beds Pliocene sandstone deposits comparable to the Diestien Beds on the French side of the Dover Strait

Miocene one of the four epochs comprising the Tertiary Period (23–5.3 million years ago)

Mollusca (molluscs) a very diverse phylum (major division) of invertebrates which have a common body plan modified in various ways

Moraine rock material carried, or having been carried and deposited, by a glacier

Newbournian formation the Red Crag in the Newbourne region of Suffolk

Norwich Crag third lowest deposit of the East Anglian Crags contains temperate climate marine fauna

Oolitic rock composed entirely or largely of small sub-spherical sand-sized carbonate particles ooliths (ooids)

Ordovician one of the six Periods comprising the Primary Era (488–444 million years ago)

Palaeometeorology study of past weather conditions

Pleistocene first of the two epochs of the Quaternary Period (1.8 million years ago to about 11,500 years ago)

Pliocene one of the four epochs comprising the Tertiary Period (5.3–1.8 million years ago)

Polyzoa phylum of minute marine animals with a calcareous skeleton, forming compound colonies

Pro-glacial lake glacial water body immediately in front or around the margin of an ice sheet

Quartzose rock mainly or entirely composed of quartz

Quaternary Period comprising the Pleistocene and Holocene Epochs (1–2 million years ago to present)

Red Crag second lowest deposit of the East Anglian Crags contains cool temperate type marine fauna including the first remains of horses and elephants

Scaldisien Pliocene deposit of Belgium; the equivalent of the Waltonian Crag

Strike horizontal line at right-angles to the slope

Tertiary Period 63–1.5 million years ago

Varve banded layer of silt and sand deposited annually in lakes, especially near to ice sheets; the coarser paler material is deposited in summer, the finer, darker material in winter; one varve comprises one light band and one dark band

Waltonian Crag lowest division of the Red Crag found in Essex

Weybourne Crag division of the Icenian Crag

Bibliography of the Writings of Frederic W. Harmer

The following list of Harmer's works (published and otherwise) illustrates the breadth of his engagement with the geology of East Anglia over the course of a long life.

Harmer, F.W. (1865). Draft paper, 'On identity of Cromer Beds and Norfolk river valleys' (BGS).

Harmer, F.W. (1866). Draft paper, 'On boulder clay in Norfolk' (BGS).

Harmer, F.W. (1867a). On the existence of a third boulder-clay in Norfolk. Quarterly Journal of the Geological Society 23: 87–90.

Harmer, F.W. (1867b). Drift of the Eastern Counties. *Geological Magazine*, **4**.

Harmer, F.W. (1869). On the so-called Crag of Belaugh and Weybourne, *Geological Magazine*, **6**.

Harmer, F.W. (1871a). Draft paper, 'On the molluscan fauna of the Norfolk coast' (BGS).

Harmer, F.W. (with S.V. Wood) (1871b). On the palaeontological aspects of the Middle Glacial Formations of East Anglia. Report British Association for 1870.

Harmer, F.W. (1877a). The Testimony of the Rocks in Norfolk. A popular description of the geology of the county. Hamilton Adams & Co., London.

Harmer, F.W. (1877b). On the Kessingland Cliff and on the relation of the Forest-Bed to the Chillesford Clay. *Quarterly Journal of the Geological Society*, **33.**

Harmer, F.W. (with S.V. Wood) (1877c). Obervations on the later Tertiary Geology of East Anglia. *Quarterly Journal of the Geological Society*, **33.**

Harmer, F.W. (with S.V. Wood) (1877d). The Kessingland Freshwater Bed and Weybourne Sand. Geological Magazine.

Harmer, F.W. (1878). Tellina Balthica and the Norwich Crag. *Transactions of the Norfolk and Norwich Naturalists' Society*, **2**.

Harmer, F.W. (1895a). The Southern Character of the Molluscan Fauna of the Coralline Crag tested by an analysis of its characteristic and abundant species. Report British Association, (Ipswich).

Harmer, F.W. (1895b). On the Derivative Shells of the Red Crag. Report of the British Association for the Advancement of Science (Ipswich).

Harmer, F.W. (1895c). Draft paper, 'On the British Geological Survey and boring of deep-seated rocks' (BGS).

Harmer, F.W. (1896a). The Pliocene deposits of Holland and their relation to the English and Belgian Crags, with a suggestion for the establishment of a new zone, 'Amstelian' and some remarks on the geographical conditions of the Pliocene epoch in Northern Europe. Quarterly Journal of the Geological Society, 52: 748–782.

Harmer, F.W. (1896b). The southern character of the molluscan fauna of the Coralline Crag. Geological Magazine, **33**.

Harmer, F.W. (ca 1896c). Draft paper, 'On Middle Glacial Sands, Coralline Crag and note from Professor Kendall' (BGS).

Harmer, F.W. (ca 1898). Typescript: 'The Lower Pliocene deposits of the Anglo-Belgian Basin' (BGS).

J. A. Kington, *Frederic W. Harmer: A Scientific Biography*, SpringerBriefs in Environment, Security, Development and Peace 19, DOI: 10.1007/978-3-319-07704-8,

Harmer, F.W. (1898). The Pliocene deposits of the east of England: The Lenham Beds and the Coralline Crag. Quarterly Journal of the Geological Society, 54: 308–356.
Harmer, F.W. (1899). Draft paper, 'On meteorological conditions of North-Western Europe during Pliocene and Glacial periods' (BGS).
Harmer, F W. (1900a). On a proposed new classification for the Pliocene deposits of the east of England. Report of the British Association for the Advancement of Science, Transactions of Section C (for 1899): 751–753.
Harmer, F.W. (1900b). The Pliocene deposits of the east of England, Part II: The Crag of Essex (Waltonian) and its relation to that of Suffolk and Norfolk. Quarterly Journal of the Geological Society, 56: 705–744.
Harmer, F.W. (1900c). The influence of the winds upon climate during past epochs; A meteorological explanation of some geological problems. Report of the British Association for the Advancement of Science, Transactions of Section C: 753–754.
Harmer, F.W. (1901). The influence of the winds upon climate during the Pleistocene Epoch: A palaeometeorological explanation of some geological problems. Quarterly Journal of the Geological Society, 57: 405–476.
Harmer, F.W. (1902). A sketch of the later Tertiary history of East Anglia. Proceedings of the Geologists' Association, 17: 416–451.
Harmer, F.W. (1903). Field excursion to Cromer, Norwich, and Lowestoft, July 9th-14th, 1903. Proceedings of the Yorkshire Geological Society, 15: 305–314.
Harmer, F.W. (1904a). Two draft papers, 'On the Great Eastern Glacier' (BGS)
Harmer, F.W. (1904b). The Great Eastern Glacier, Spottiswood, London (reprinted from the Report of the British Association for the Advancement of Science, *Transactions of Section C*: 542–543.
Harmer, F.W. (1905). The Great Eastern Glacier. Report of the British Association for the Advancement of Science (Cambridge).
Harmer, F.W. (1906a). Draft paper, 'On the Weybourne zone, East Anglia' (BGS)
Harmer, F.W. (1906b). Draft paper, 'The glacial deposits of the East of England' (BGS)
Harmer, F.W. (1906c). The Glacial Deposits of the East of England. Report of the British Association for the Advancement of Science (York).
Harmer, F.W. (1907). On the origin of certain cañon-like valleys associated with lake-like areas of depression. Quarterly Journal of the Geological Society, 63: 470–514.
Harmer, F.W. (1910a). The glacial deposits of Norfolk and Suffolk. Transactions of the Norfolk and Norwich Naturalists' Society, 9: 108–133.
Harmer, F.W. (1910b). The Pliocene Deposits of the Eastern Counties of England. In: *Geology in the Field:* The Jubilee Volume of the Geologists' Association (1858–1908), edited by H.W. Monckton and R.S. Herries, Part I, 86–102, Edward Stanford, London.
Harmer, F.W. (1910c). The Pleistocene Period in the Eastern Counties of England. In: *Geology in the Field:* The Jubilee Volume of the Geologists' Association (1858–1908), edited by H.W. Monckton and R.S. Herries, Part I, 103–123, Edward Stanford, London.
Harmer, F.W. (1914–1919). *A Monograph of the Pliocene Mollusca of Great Britain*, being supplementary to Searles V. Wood's *Monograph of the Crag Mollusca*, Volume I, Palaeontographical Society, London.
Harmer, F.W. (1920–1925). *A Monograph of the Pliocene Mollusca of Great Britain*, being supplementary to Searles V. Wood's *Monograph of the Crag Mollusca*, Volume II, Palaeontographical Society, London.
Harmer, F.W. (1925) (posthumously). Further remarks on the meteorological conditions of the Pleistocene Epoch, Quarterly Journal of the Royal Meteorological Society, 51: 247–259.
Harmer, F.W. (1928) (posthumously). The distribution of erratics and drift. Proceedings of the Yorkshire Geological Society, 21: 79–150.
Harmer, F.W. (nd). Numerous papers on a geological history of Norfolk (BGS).
Harmer, F.W. (nd). Pages of a paper: 'A sketch of the geological history of Norfolk' (BGS)
Harmer, F.W. (nd). Paper: 'On Geology of Norfolk' (BGS).

Harmer, F.W. (nd). Draft notes referring to Boulder Clay in Norfolk (BGS).
Harmer, F.W. (nd). Draft paper, 'Popular notes on the geology of Norfolk' (BGS).
Harmer, F.W. (nd). Draft paper, 'On Crag of Belaugh & Weybourne' (BGS).
Harmer, F.W. (nd). Draft paper, 'On Lake Oxford and the Goring Gap' (BGS).
Harmer, F.W. (nd). Draft paper, 'On Pliocene deposits of Tegelen and Forest-Bed Cromer (BGS).
Harmer, F.W. (nd). Draft paper, 'On the origin of certain canon-like valleys associated with lake-like areas of depression' (BGS).
Harmer, F.W. (nd). Draft paper, 'On the Lowestoft boring' (BGS).
Harmer, F.W. (ca 1907). Photographs mainly of Essex, Suffolk and Norfolk [IGS1/733] (BGS).
Harmer, F.W. (1911). Photograph album, Norfolk. Inscribed 'A souvenir of a beautiful excursion in Norfolk given to his excellent guide Mr Harmer. Albrecht Penck' [IGS1/733] (BGS).
See also the special website on this book on F.W. Harmer: http://afes-press-books.de/html/SpringerBriefs_ESDP_Harmer.htm

About the Author

Born in London, 1930, John A. Kington joined the U.K. Meteorological Office in 1947 gaining experience in synoptic meteorology as a weather observer and forecaster both at home and abroad. In 1958, he obtained his first degree, B.Sc. in Geography, at the University of London, studying under Professor Gordon Manley and in 1959 passed the Advanced Forecasting Course of the Meteorological Office. After gaining his M.Sc. in Meteorology in 1969 he began his academic career with the study of Climatic Change and Historical Climatology, first at the University College of Swansea and, since 1971, with Professor Hubert Lamb at the Climatic Research Unit, School of Environmental Sciences, University of East Anglia in Norwich, where he is Visiting Fellow. In 1999, he was awarded the Jehuda Neumann Memorial Prize of the Royal Meteorological Society.

Among his major research interests are: *Historical Climatology*: *The Weather of the 1780s over Europe*, Cambridge University Press, 1988; digitally printed version, 2009; *The Weather Journals of a Rutland Squire: Thomas Barker of Lyndon Hall*, Rutland Record Society, 1988; *Climate and Weather*, Collins New Naturalist Library, HarperCollins, London, 2010. *Synoptic Meteorology*: *The Weather Book*, Michael Joseph, 1982; *Even The Birds Were Walking: The Story of Wartime Meteorological Reconnaissance*, Tempus, Stroud UK and Charleston, USA, 2000; *WEKUSTA: Luftwaffe Meteorological Reconnaissance Units & Operations 1938–1945*, Flight Recorder Publications, 2007; *Research Projects: Atlantis der Nordsee* [Atlantis of the North Sea], A film by Gabriele Wengler, ECO Media TV-Produktion GmbH, Hamburg 2010; An Oral history of British Science: John Kington: Meteorologist-Climatologist (interview by Paul Merchant), National Life Stories, British Library Sound Archive, London, October 2010–February 2011. A list of his over fifty articles published in various scientific journals can be found in the website of this book at: http://afes-press-books.de/html/SpringerBriefs_ESDP_Harmer.htm.

J. A. Kington, *Frederic W. Harmer: A Scientific Biography*, SpringerBriefs in Environment, Security, Development and Peace 19, DOI: 10.1007/978-3-319-07704-8,

About the Book

Comprising the first definitive account of the geological and palaeometeorological studies made by Frederic W. Harmer (1835–1924), this book contributes a previously missing chapter to the history of science. The main objective of the author is to ensure that the scientific work of Harmer, which unfortunately has been widely neglected or forgotten, becomes more generally known and acknowledged. The balance of this deficiency will be redressed by bringing to light in this volume his contributions to the history of science to an audience of academic and lay readers of the current literature.

Harmer was one of the pioneers in the field of East Anglian geology as well as one of the last members of a distinguished group of amateur geologists who had been responsible for making major advances in the science during the Victorian era and early years of the 20th century. In particular, he played a key role in elucidating the Pliocene and Pleistocene stratigraphy in the east of England by developing the use of mollusca for biostratigraphic correlation within the Crags of East Anglia.

He was certainly a well-respected scientist in his day, being awarded Hon. M.A. Cantab., elected F.G.S. and F.R.Met.S., and a Membre Hon. de la Société Belge de Géologie et de Paléontologie. From an early age he had been an active member of the Geological Society of Norwich. In 1902 he was awarded the Murchison Medal for that year in recognition of his work on the Pliocene and Pleistocene deposits of East Anglia. A leading exponent of 19th-century geological literature, Horace B. Woodward, F.R.S., F.G.S., included Harmer in a group of geologists who 'have added greatly to our knowledge of the structure of the [East Anglian Crag] deposits'.

J. A. Kington, *Frederic W. Harmer: A Scientific Biography*, SpringerBriefs in Environment, Security, Development and Peace 19, DOI: 10.1007/978-3-319-07704-8,

MIX
Papier aus verantwortungsvollen Quellen
Paper from responsible sources
FSC® C105338

If you have any concerns about our products,
you can contact us on
ProductSafety@springernature.com

In case Publisher is established outside the EU,
the EU authorized representative is:
Springer Nature Customer Service Center GmbH
Europaplatz 3, 69115 Heidelberg, Germany

Printed by Libri Plureos GmbH
in Hamburg, Germany